FORCED HOT AIR FURNACES
TROUBLESHOOTING AND REPAIR

Roger Vizi

McGraw-Hill

New York   San Francisco   Washington, D.C.   Auckland   Bogotá
Caracas   Lisbon   London   Madrid   Mexico City   Milan
Montreal   New Delhi   San Juan   Singapore
Sydney   Tokyo   Toronto

Library of Congress Cataloging-in-Publication Data

Vizi, Roger.
    Forced hot air furnaces : troubleshooting and repair / Roger Vizi.
      p. cm.
    Includes index.
    ISBN 0-07-134171-4
    1. Hot-air heating—Equipment and supplies—Maintenance and
repair. I. Title.
    TH7601.V57 1999
    697'.3—dc21   99-18119
                                                 CIP

McGraw-Hill

A Division of The McGraw·Hill Companies

1 2 3 4 5 6 7 8 9 0 DOC/DOC 9 0 4 3 2 1 0 9

ISBN 0-07-134171-4

The sponsoring editor for this book was Zoe G. Foundotos, the editing supervisor was Paul R. Sobel, and the production supervisor was Pamela A. Pelton. It was set in Melior per the CMS design specs by Michele Zito and Paul Scozzari of McGraw-Hill's Professional Book Group Hightstown composition unit.

Printed and bound by R. R. Donnelley & Sons Company.

McGraw-Hill books are available at special quantity discounts to use as premiums and sales promotions, or for use in corporate training programs. For more information, please write to the Director of Special Sales, McGraw-Hill, 11 West 19th Street, New York, NY 10011. Or contact your local bookstore.

This book is printed on recycled, acid-free paper containing a minimum of 50% recycled de-inked fiber.

CONTENTS

ACKNOWLEDGMENTS

I would like to take the time to thank several people and organizations who assisted me in the preparation of this book. First, to my wife, Nadine, who allowed me the time to write, and encouraged me in this endeavor. To my father, Edward Vizi, for his encouragement. To my father-in-law, Robert Speers, for the testing of the original version of this book.

The following companies and organizations contributed to this book: Lennox Industries, The Honeywell Company, The U.S. Department of Energy, Energy Efficiency and Renewable Energy Clearinghouse, Mr. Michael Lamb (1-800-237-2957). Illustrations by Wade Owens.

FORCED HOT AIR FURNACES

Introduction

As a heating professional, you will be called on to service many types of home heating systems. This book will cover the most common types of home heating systems in use today: natural and propane gas, oil, electric, and heat pumps. These systems come in many styles and are manufactured by several companies, but the basic operation and repair are the same.

While I have geared this book for the person who is either in the heating profession now or would like to become a heating professional, I also have made every attempt to make this material as easy to understand as possible for the average homeowner. While the average homeowner may never attempt to replace the heat exchanger in his or her own unit, as an example, he or she will have the knowledge to decide when this type of repair is needed.

Wherever I felt that more information on a subject was needed beyond the discussion included herein, I have included diagrams and drawings to better illustrate the point.

As a heating system professional, you must have a complete understanding of the different types of add-ons that are available for home heating systems as well. I have included sections on installation and maintenance of the two most common types of humidifiers on the market, when you should install one, which one is right for the application, and how to properly maintain these units. I also explain the benefit to some homeowners of installing an electronic air cleaner.

Each section of this book will cover the basic controls of the heating system and the operation of the system and will end with a discussion of troubleshooting and repair of the system.

It is important for the heating professional to follow a systematic approach to diagnosing problems when they occur. By understanding the different components of the heating system and how they work together, you can isolate the problem to a particular circuit and eliminate the circuits that do not have a bearing on the problem.

You will learn, for example, that if a client calls you and states that the burners will not light but the blower runs all the time that these two conditions have no bearing on each other. Both these problems are controlled by different circuits, and the problems are not related. However, by knowing the proper troubleshooting techniques, you should begin to diagnose the problem in your mind so that you have an idea of what to look for when you arrive at the client's home.

Chapter 2 of this book will go into more detail on the subject of listening and observing. These are two of the most important traits that a good heating professional can have, since the client is your best source of information on the problem.

The working world of the heating professional is a rewarding and challenging one. You will encounter many different situations that you must have the knowledge and experience to handle. While this book does not attempt to cover all the situations that you may be called on to handle, it will explain the techniques used to troubleshoot and repair the most common problems. By applying these same techniques to your particular challenge, you should be able to come up with a solution quickly and efficiently. This knowledge and confidence will be apparent to your clients, and they will call on you again because you have shown them that you are truly a professional at your trade.

Listening and Observing

One of the most important parts of the job of heating professional is knowing what to look for when you receive a call from a homeowner that his or her gas heating system, as an example, is in need of repair. You must learn how to listen to the homeowner, because he or she is your most valuable source of information.

The homeowner may tell you, "I do not have any heat in the house." It is your job to be able to ask the proper questions that will lead you to the correct diagnosis of the problem.

When you receive this type of call, you should ask the following questions:

1. Is the thermostat turned up?

2. Did you check the fuses?

3. Is the pilot light lit?

4. Did you hear any noise when the heating system was running?

5. Were any repairs made in the last several months?

6. What type of heating system do you have? If this is an oil or propane system, ask this question:

7. When was the last time fuel was delivered?

All these questions will help you to diagnose the problem before you leave for the call (see Appendix B). Even though the client has answered all these questions and has assured you that all these items have been checked, it is always good practice to double-check this entire list once you arrive at the home. Some homeowners are not aware of the location of all the fuses or breakers that are attached to the circuits, for example, and you may find that this is the answer to the problem. If you do not double-check these items, you may find yourself looking for a ghost.

In the chapters on troubleshooting, I will cover many of the techniques that are required to properly diagnose the symptoms and to correct the problems of a heating system that is not operating properly. As you begin to use these techniques, you will find that it will become much easier to solve these problems in a timely manner. The more troubleshooting calls that you go on, the more you will understand why it is so important to listen to the homeowner, ask the proper questions, and observe what is taking place with the heating system so that you can quickly diagnose the problem and make the needed repairs.

You should always take a systematic approach to troubleshooting any heating problem. Always check for the simple solution to the problem first before proceeding. By using this approach, you will solve more problems quickly and will not waste time "chasing your tail" because you are looking at a symptom and not the problem.

Components of a Gas Forced Air Heating System

Before we can begin to understand how to repair common gas furnace problems, we must first get an understanding of what makes up a home gas heating system and how the components work together. The typical home gas heating system is simply a series of switches and circuits that work together to heat the home. As we begin to explain each component, it would be helpful to locate each component described so that when a home heating repair is needed on a gas heating system, you are already familiar with the location of these components. Figure 3.1 shows a typical gas forced air heating system.

Thermostat

The *thermostat* is the device used to regulate the temperature in the home. It is the "switch" that tells the home gas heating system

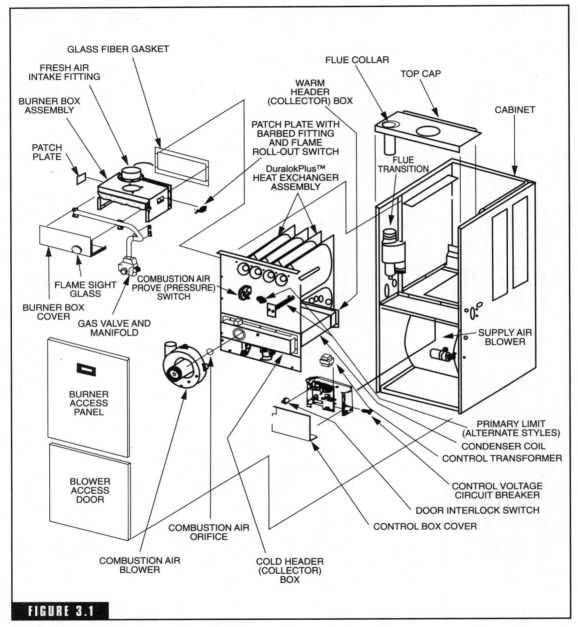

FIGURE 3.1

Exploded view of gas forced air heating system. (*Courtesy of Lennox Industries, Inc.*)

to begin the heating cycle. This device comes in several styles and typically is installed on an interior wall of the home. Some of these styles include round, square, and rectangular. All these thermostats operate in the same basic manner. Figures 3.2 to 3.4 show these thermostats.

When the dial or lever is moved to the right and is pointing to a temperature that is higher than the temperature in the home, the circuit is closed, and a signal is sent to the gas valve to begin the heating cycle. Some of these thermostats have one or two levers that are manual controls. One of these levers will say, "Heat Off Cool," and the other will say, "Fan On Off." It is important to know about the location of these levers and what they control. On occasion you may have to diagnose a heating problem that is related to one of these switches being placed in the wrong position.

Another part of the thermostat is called the *heat anticipator.* This is a small coil that is heated. It is used to anticipate when the temperature in the home is reaching the set point on the thermostat. Figure 3.5 shows a heat anticipator on a thermostat. When the temperature is getting close to the set point, the thermostat will shut off the call for heat to the gas valve to allow the blower to remove the balance of heat in the system so that the actual temperature in the room is correct. If this device is not set properly, the heating system will not be synchronized, and

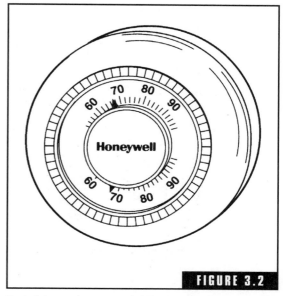

FIGURE 3.2

Round-type thermostat. (*Courtesy of Honeywell, Inc.*)

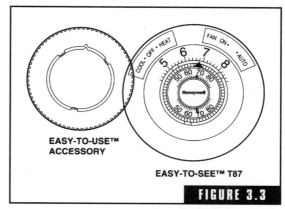

FIGURE 3.3

Round thermostat with heat/cool/fan control. (*Courtesy of Honeywell, Inc.*)

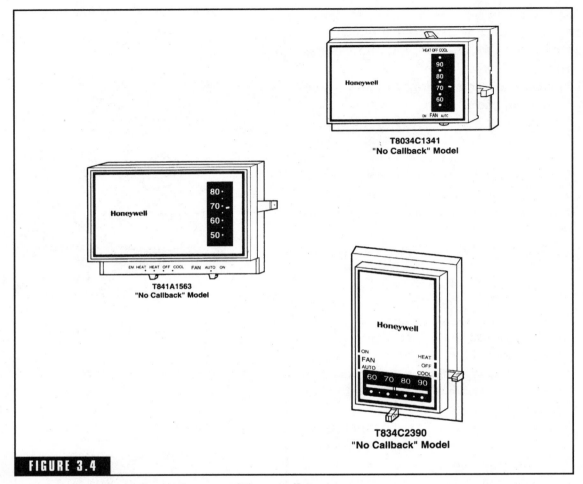

FIGURE 3.4

Various styles of thermostats. (*Courtesy of Honeywell, Inc.*)

the burner may turn on and off erratically. It is scaled in amps, and a good starting point is 0.4 to 0.5.

Gas Valve

The *gas valve* is the device that controls the natural gas flow (gas supplied by the gas company in your area) or propane gas flow (held in a tank outside the home) to the home heating system when the thermostat calls for heat. This device typically is located behind the front panel of the gas furnace. This location could be either in the upper section on an upflow-type furnace or in the lower section on a downflow-type furnace. Figures 3.6 through 3.8 show combination, continuous pilot, and electronic gas valves.

There are many other possible locations for this device depending on the type of gas furnace you have. In most cases, this device can be located by following the large gas line to the furnace. The gas valve will be either round or square and will have wires attached to it as well as typically two smaller lines coming from it. One may be silver in color and the other gold. Most will have a dial on the top that reads on, off, and pilot. The silver and gold lines lead to the pilot assembly.

Pilot Assembly

The *pilot assembly* consists of the pilot light and the thermocouple on older units and a pilot, spark ignition, and sensor on newer units. Both these types of units typically will be attached to the gas valve and will be located by the main burner(s). There may be an access panel covering this unit that must be removed or lifted to gain access to the pilot assembly. In some instances, especially on older gas furnaces, this may not be the case. In an older gas furnace, the "pilot gas line" may be attached to the main supply line by means of a small brass petcock valve, and the thermocouple may be attached to a

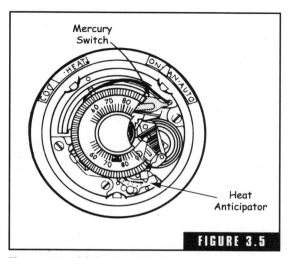

FIGURE 3.5

Mercury Switch

Heat Anticipator

Thermostat with heat anticipator.

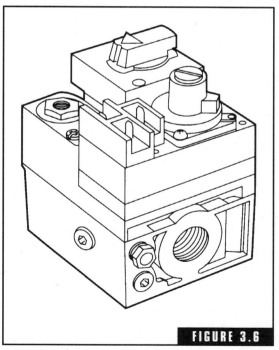

FIGURE 3.6

Combination gas valve. (*Courtesy of Honeywell, Inc.*)

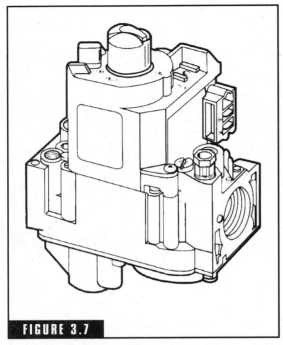

FIGURE 3.7

Continuous pilot dual automatic gas valve. (*Courtesy of Honeywell, Inc.*)

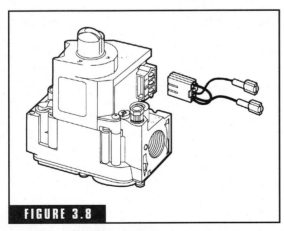

FIGURE 3.8

Universal electronic ignition gas valve. (*Courtesy of Honeywell, Inc.*)

silver box attached to the inside of the gas furnace behind the access door. Figures 3.9 and 3.10 show these types of pilot assemblies.

The main purpose of the pilot assembly is to maintain a steady pilot flame to ignite the burner(s) in the gas heating system when there is a call for heat. The job of the thermocouple is to sense that the pilot light is burning prior to the gas valve opening to supply the main gas supply to the burner(s). If the thermocouple does not sense that the pilot flame burning, it will shut off the supply of gas to the pilot and not allow the main valve to open. It is one of several safety devices on your home gas furnace. Figure 3.11 shows a thermocouple.

By keeping this pilot lit during the non-heating-season months, it will not allow moisture to form causing rust that can reduce the efficiency and life of your home heating system.

Fan and Limit Control

This is a device that controls when and how long the blower on your furnace will remain on to force heat into your home. (This is where the term *forced air gas heat* comes from). Figure 3.12 shows a fan and limit control.

This device typically is located on the front of the furnace above the burner(s) or may be located on the hot air plenum (the sheet metal that is attached to the heating section of the furnace, better known as the *supply side*). This device also serves as another form of safety device; if the blower does not come on during a heating cycle, this device will shut off the flow of gas to the burners to help eliminate the chance of overheating your furnace.

The most common types of these controls are as follows:

1. *Dial type.* In this type, there is a dial on the front of the fan control that allows you to set the on and off temperature of the blower. This type of fan control usually will have a manual blower switch attached so that the homeowner can run the blower manually in the summer if needed.

2. *Rectangular type.* This is a bimetallic type of fan control that has slide levers that are used to control the on and off settings of the blower. This style of control will control the blower's manual setting by moving the lever all the way to the left.

3. *Clicson type.* This type uses a disk that has preset on and off settings. You can not set the manual control of the blower with this type.

Blower Assembly

The *blower assembly* is the device that blows the warm air into your home during the heating cycle. This device is

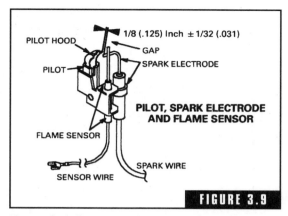

Electronic pilot assembly. (*Courtesy of Lennox Industries, Inc.*)

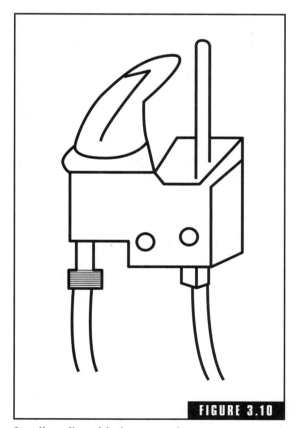

Standing pilot with thermocouple.

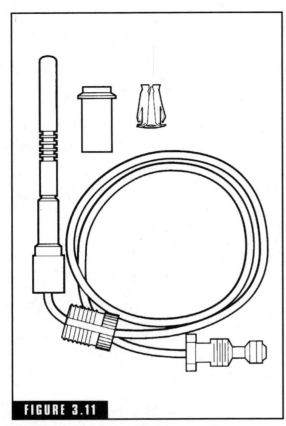

FIGURE 3.11

Thermocouple. (*Courtesy of Honeywell, Inc.*)

made up of the blower housing and a motor. The motor will be connected to the blower housing in one of two ways. The motor may be attached by means of a belt and pulley; this is known as *belt drive* (Fig. 3.13). Alternatively, the blower may be attached directly to the motor shaft; this is known as *direct drive* (Fig. 3.14). In either case, the basic function is still the same.

The belt-drive method requires more maintenance than the direct-drive method and will be explained further in the section entitled, Summer Tune-up. The blower is controlled during the heating cycle by the fan control.

Heat Exchanger

The *heat exchanger* is the unit that the burners are attached to (Fig. 3.15). When the burners come on to start the heating cycle, the heat exchanger performs two functions:

1. It conducts the heat transfer from the burners, to cause the fan control to start the blower, forcing warm air into the home.

2. It is the channel for the removal of carbon monoxide to the chimney.

This second item is extremely important to know because a heat exchanger that is damaged or not in proper operating condition may allow carbon monoxide to enter the home. Carbon monoxide is a colorless, odorless, tasteless gas that can cause the homeowner and his or her family to become very sick, and in high enough concentrations, death can occur. How to examine this vital part of the heating system will be covered in Chapter 5, Tuning Up a Gas Forced Air Heating System.

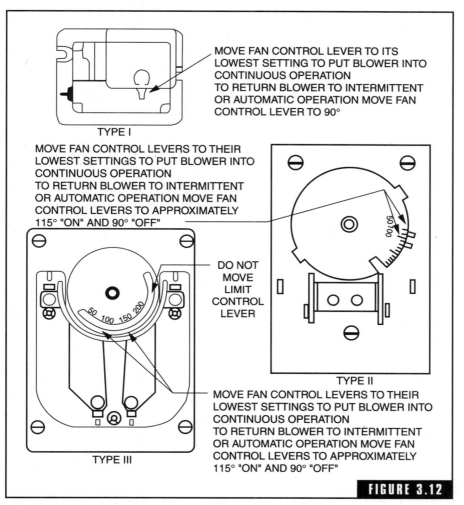

MOVE FAN CONTROL LEVER TO ITS
LOWEST SETTING TO PUT BLOWER INTO
CONTINUOUS OPERATION
TO RETURN BLOWER TO INTERMITTENT
OR AUTOMATIC OPERATION MOVE FAN
CONTROL LEVER TO 90°

TYPE I

MOVE FAN CONTROL LEVERS TO THEIR
LOWEST SETTINGS TO PUT BLOWER INTO
CONTINUOUS OPERATION
TO RETURN BLOWER TO INTERMITTENT
OR AUTOMATIC OPERATION MOVE FAN
CONTROL LEVERS TO APPROXIMATELY
115° "ON" AND 90° "OFF"

DO NOT
MOVE
LIMIT
CONTROL
LEVER

TYPE II

TYPE III

MOVE FAN CONTROL LEVERS TO THEIR
LOWEST SETTINGS TO PUT BLOWER INTO
CONTINUOUS OPERATION
TO RETURN BLOWER TO INTERMITTENT
OR AUTOMATIC OPERATION MOVE FAN
CONTROL LEVERS TO APPROXIMATELY
115° "ON" AND 90° "OFF"

FIGURE 3.12

Different types of fan/limit controls. (*Courtesy of Lennox Industries, Inc.*)

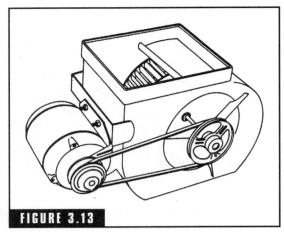

Belt-drive blower unit.

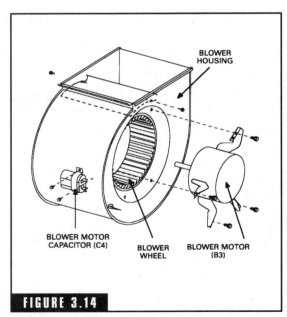

BLOWER
HOUSING

BLOWER MOTOR
CAPACITOR (C4)

BLOWER
WHEEL

BLOWER MOTOR
(B3)

Exploded view of a direct-drive blower assembly.
(*Courtesy of Lennox Industries, Inc.*)

Gas Regulators

A *gas regulator* is the device that controls the amount and pressure of the gas that is used for combustion in the gas heating system. In the case of liquid propane (LP) gas (also known as *propane gas*), the regulator is in the line between the tank and the gas valve. With propane, there will be no regulator on the gas valve. In the case or natural gas, the regulator will be located on the gas valve. This is one quick way for the heating professional to determine which type of fuel is being used.

Hot Air Ducting

This is also known as the *supply-side ducting*. This is the ducting that comes from the hot air side of the heating system and supplies the warm air to the home. This ducting may be round or square and can be the flex type of metal. Figure 3.16 shows different types of ducting. The outlets for this warm air are registers that should be located along the exterior walls of the home. They can be used to regulate the amount of heat that is provided to each room of the home. A later chapter will describe how to balance the heating system by regulating these registers.

Cold Air Ducting

This is also known as *return ducting*. This is the metal ducting that returns the cold air from the home to the blower side of the heating system. The registers for this ducting should be located along the interior walls of the home. It is important to note that you cannot warm a home if you cannot properly remove the cold air from the home. One cold air return is worth two warm air registers in a home. If you do not see enough cold air return registers in a home, you should talk to the homeowner to see if there are "cold spots" in the home in the winter. If the answer is yes, you should recommend adding more cold air returns.

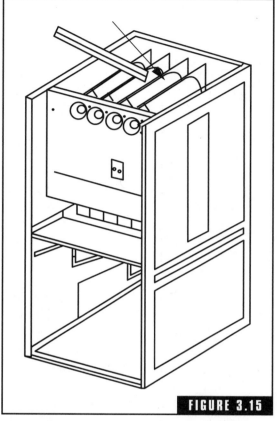

FIGURE 3.15

Heat exchanger. (*Courtesy of Lennox Industries, Inc.*)

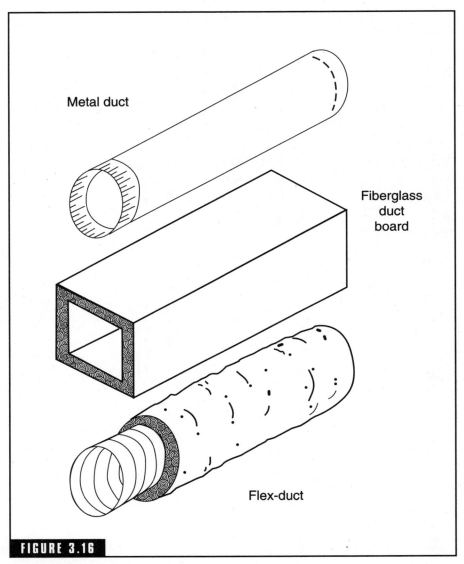

Metal duct

Fiberglass
duct
board

Flex-duct

FIGURE 3.16

Types of ducting.

Electric Circuits

All the heating systems covered in this book use electricity for power. Some of the units use this power to operate the electric circuits, others use it to generate heat, and heat pumps use it to move heat from one place to another. As a heating professional, it is important for you to understand how these circuits operate so that you are better equipped to troubleshoot these circuits.

I will start out with an explanation of the circuits used in the gas forced air heating system. As I mentioned earlier, a heating system uses a series of switches to control the operation of the system. These switches use either low voltage (24 V ac) or a higher voltage (110 V ac) to operate.

Low-voltage circuits primarily operate the thermostat and some of the gas-valve circuits. A *stepdown transformer* is used to convert the 110 to 24 V ac. In some installations, the thermostat will be connected to the 110 V ac circuit. You can tell this by looking at the size of the wire that is used. On a 24-V circuit, the wire is very thin and normally will be two colors, red and black. The wire will be single solid copper wire. On the 110-V circuit, the wire will be much larger in diameter, and the colors may be black and white. Of course, you should always check the voltage with a voltmeter to make sure of the circuit with which you are working.

Let's first examine the low-voltage circuit of a gas heating system. The stepdown transformer typically will be located inside the wiring cabinet. Figure 4.1 shows a typical 24-V transformer. This cabinet is located in the blower section of the heating system. Some other locations for this transformer are in the front of a "low boy" heating system or, on older systems, on the outside of the cabinet or mounted somewhere close to the heating system.

There are two wires that run from the back of the transformer, one white wire and one black wire. The black wire is considered the "hot" wire, and the white wire is the ground wire. The transformer will be wired into the constant 110-V power supply that is being fed in from the SPST (single pole single throw) switch (the switch on the outside of the heating system that has a switch and fuse, Fig. 4.2) or from the breaker or fuse panel (Fig. 4.3). With the transformer mounted and wired properly, you will be able to get a reading of 24 V from the two screws on the front of the transformer. Using a voltmeter set for 24 V ac, place the black lead on one of the screws and the other lead on a grounding source (bare metal, etc.). You should get a reading. If not, try moving the black lead to the other screw. The screw that produces the 24-V reading is the hot side; the other is the ground.

As the 110-V power enters the transformer, it goes through a stepdown coil, which converts the power to 24 V ac. It is very important for you as a heating technician to understand how this power conversion works.

The thermostat is wired in the 24-V circuit to act as the main switch that allows

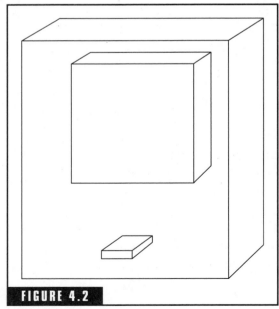

FIGURE 4.1

A 24-V transformer and blower relay. (*Courtesy of Lennox Industries, Inc.*)

FIGURE 4.2

SPST switch.

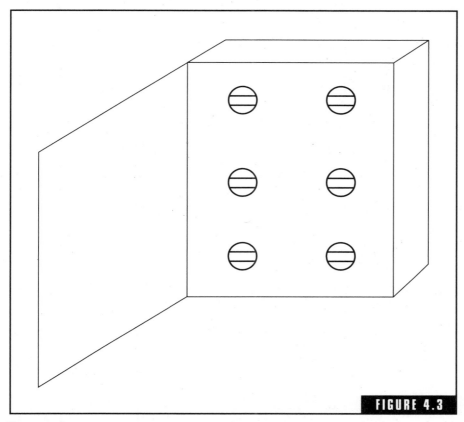

FIGURE 4.3

Fuse panel.

the gas valve to open when the thermostat is turned up and is calling for heat. The hot wire from the transformer is run to one side of the thermostat. Another wire is run from the other terminal on the thermostat the one terminal on the low-voltage side of the gas valve. The other wire is connected between the other terminal on the low-voltage side of the gas valve and then back to the other terminal on the transformer. Because the hot side of the transformer voltage is "broken" at the thermostat, when the thermostat is turned up and is calling for heat, this closes the circuit and allows the voltage to flow to the gas valve. This, in turn, energizes a coil that opens the flow of main supply gas to the heating system, starting the heating cycle.

On the high-voltage side (110 V ac), the power is supplied directly from the main fuse box or breaker panel. This power normally will be

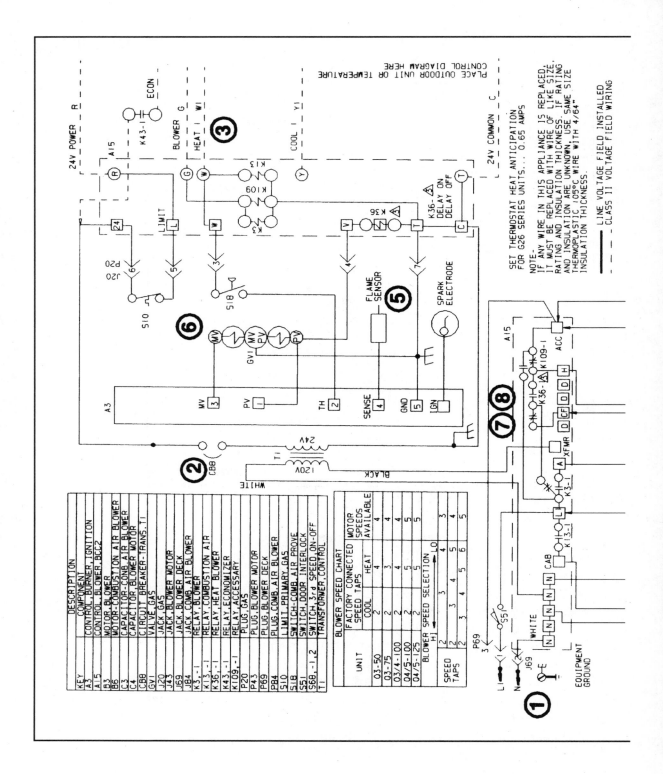

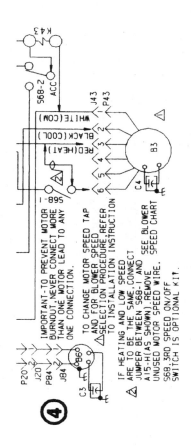

E– G26-1 and -2 models.

1. When disconnect is closed, 120V is routed through door interlock switch (S51) to feed the line voltage side of the blower control (A3) and the Transformer T1 primary. Door interlock sxitch must be closed for A3 and T1 to receive voltage.

2. T1 supplies 24VAC to terminal "24VAC" on A3. In turn, terminal "R" of A3 supplies 24VAC to terminal "RC" of the indoor thermostat (not shown).

3. When there is a call for heat, W1 of the thermostat energizes W of the furnace control with 24VAC.

4. CAB of the blower control energizes the combustion air blower (B6). When the combustion air blower nears full speed, combustion air prove switch (S18) closes.

5. When S18 closes, assuming primary limit (S10) is closed, the ignition control opens the pilot valve and begins spark.

6. When flame is sensed, spark stops and main valve opens to light main burners.

7. After 45 seconds, blower control (A3) energizes the indoor blower.

8. When heat demand is satisfied, W1 of the thermostat de-energizes W of the furnace control and the furnace control immediately de-energizes the gas valve. The combustion air blower immediately stops. Also, the indoor blower runs for a designated period (90–330 seconds) as set by jumper on blower control.

FIGURE 4.4

Wiring diagram for spark ignition system. (*Courtesy of Lennox Industries, Inc.*)

connected to a switch with a fuse located on the outside of the heating unit.

The blower circuit is then connected to these wires. The black wire is the hot wire, and the white wire is the ground. There also may be a bare copper wire that is also a grounding wire. You must make sure that you always connect the same color wires together. Never connect the white (ground) wire to the black (hot) wire or you will blow a fuse or worse.

In this circuit, the fan control is acting as the switch. It is connected into the circuit so that when the burners are on and the fan control reaches the desired on set point, the circuit is closed, and power is sent to the blower to operate. This same circuit is used to send power to the high-voltage side of the gas valve. In the event that the blower does not come on for any reason, the circuit will open and will close the flow of main gas supply to the burners. This is a safety device that is built in so that the unit will not overheat. Figure 3.12 shows how this circuit operates.

By understanding how these circuits operate, you will be able to better isolate the problem when you are called on to troubleshoot a gas heating system. By knowing that the low-voltage circuit only operates the thermostat and low-voltage side of the gas valve, as an example, you can eliminate the circuit if the problem involves the blower not operating. This knowledge will save valuable troubleshooting time.

Figure 4.4 shows a typical wiring diagram for a spark ignition gas forced air heating system.

Operation of a Gas Forced Air Heating System

There are three types of gas forced air heating systems that you will encounter as a heating professional. These three types are:

1. *Standing pilot systems.* In this type of system, the thermocouple is used as a safety device to detect that the pilot is burning prior to the gas valve opening to supply main-line gas to the burners. Figure 5.1 shows a thermocouple.

2. *Thermopile generator system.* The thermopile generator uses the heat from the pilot to operate the gas valve. No outside power source is needed in this type of unit. Figure 5.2 shows a thermopile generator.

3. *Electronic spark ignition system.* When the thermostat calls for heat, an electronic circuit is closed, causing a high-energy spark to start. Once this spark is detected, the pilot side of the gas valve opens to light the pilot. A sensor is used to detect the pilot and sends a signal to the gas valve to open, allowing main-line gas to

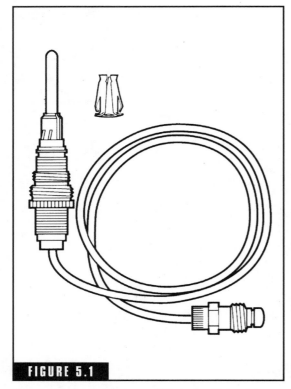

FIGURE 5.1

Thermocouple. (*Courtesy of Honeywell, Inc.*)

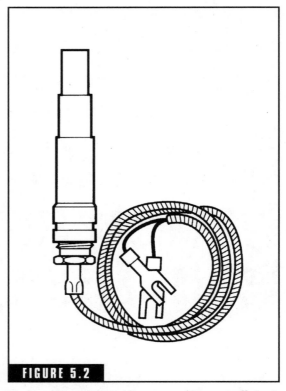

FIGURE 5.2

Thermopile generator. (*Courtesy of Honeywell, Inc.*)

flow to the burners, starting the heating cycle. Figure 5.3 shows this type of system.

There are other types of systems on the market as well, but they all work in one of these three basic ways.

In a standing pilot system, the gas forced air heating system operates in the following manner:

1. The thermostat is turned up or begins to call for heat.

2. The circuit is closed at the thermostat, and this allows electricity to flow to the gas valve.

3. The gas valve checks to make sure that the pilot is lit by means of the thermocouple. If the thermocouple senses the pilot, the gas valve opens, allowing main-line gas to flow, and the pilot ignites the burners.

4. Once the temperature reaches the on set point of the fan control, the blower begins to run, forcing warm air into the home.

5. This operation continues until the thermostat reaches its set point, and the circuit is opened, closing the main gas supply to the burners.

6. The blower continues to operate until the lower limit on the fan control is reached. This opens the circuit, and the blower stops.

A gas forced air heating system that is equipped with electronic ignition is slightly different and operates as follows:

1. The thermostat is turned up or calls for heat. This closes the circuit and sends a signal to the pressure switch to begin the 15-s purge cycle.

2. The system then begins the 20-s igniter warm-up cycle.

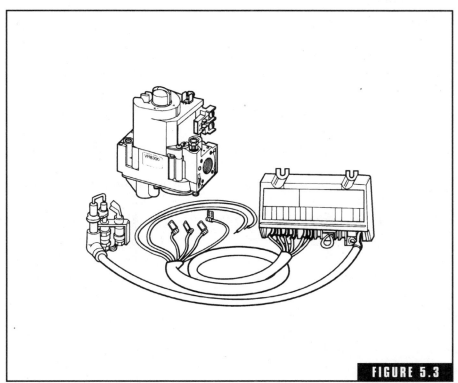

FIGURE 5.3

Components of a spark ignition system. (*Courtesy of Honeywell, Inc.*)

3. The igniter begins a 4-s trial for ignition.

4. The burners ignite, and the flame sensor checks for flame.

5. The blower delay sequence begins.

6. The blower begins the forced air cycle.

7. Once the thermostat is satisfied, the combustion blower continues 5 s postpurge.

8. The main blower runs until the low limit is reached and then shuts down.

No matter which type of gas heating system you are called to work on, the basic operation is the same. Figure 5.4 shows the sequence of operation of a spark ignition system.

All the newer home heating systems are designed for maximum efficiency and fuel savings. To accomplish this, many manufacturers have incorporated more efficient heat exchangers, burners, ignition systems, controls, etc. One such device is the automatic vent damper. This device is installed in the flue and is used to close off the flue pipe when the heating cycle is complete to allow the unit to be more efficient.

When the unit calls for heat, a signal is sent to the damper to open. Once the damper opens, the normal cycle begins. This type of device can be used on any type of gas forced air heating system. Figure 5.5 shows such a damper.

I will now explain how all this comes together to produce heat for the home. In a forced air heating system, fossil fuel is burned to produce the heat. Air enters the forced air heating system to provide oxygen to the burners, and the combustion products are vented to outdoors in the combustion airstream, usually through the chimney (Fig. 5.6). The combustion airstream moves because the combustion air products are lighter than the cold air. Movement is sometimes also assisted by a combustion air intake fan.

Another airstream moves from the return air grill, through the return air ducts and filter, to the blower, which pushes the air past the heat exchanger (Fig. 5.7). The circulated airstream and the combustion airstream are separated at the heat exchanger and not allowed to mix, as seen in Fig. 5.8. The circulated air is heated as it passes the heat exchanger. From there, the heated air passes through the supply ducts

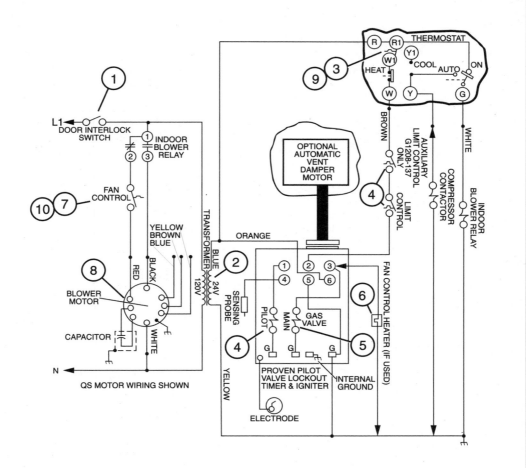

1 - Line potential feeds through the door interlock (if used). The blower access panel must be in place to energize machine.
2 - Transformer provides 24 volt control circuit.
3 - On a heating demand the thermostat heating bulb makes
4 - The control circuit feeds from "W" leg through limit control(s) to initiate pilot operation.
5 - After the pilot flame has proven, the main valve is energized. Main burners are ignited.
6 - As the main valve is energized, the fan control heater (if used) is also activated.
7 - After a short period, the heater provides sufficient heat to close the fan control contacts.
8 - This then energizes the blower motor on low speed.
9 - As the heating demand is satisfied, the thermostat heating bulb breaks. This de-energizes the ignition control, gas valve and fan control heater.
10 - The blower motor continues running until the furnace temperature drops below fan control set point.

FIGURE 5.4

Sequence of operation. (*Courtesy of Lennox Industries, Inc.*)

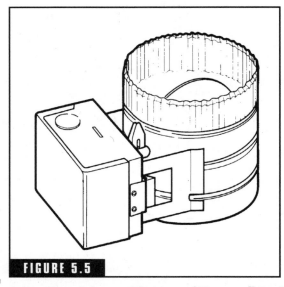

FIGURE 5.5

Automatic vent damper. (*Courtesy of Honeywell, Inc.*)

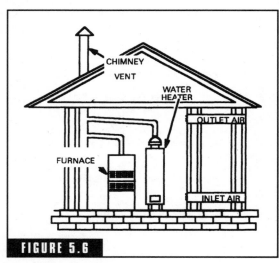

CHIMNEY
VENT

WATER
HEATER

OUTLET AIR

FURNACE

INLET AIR

FIGURE 5.6

Chimney vent system.

and past dampers, which are used to balance the airflow in the home, and then through supply diffusers and into each room of the home.

At the heart of the forced air heating system is the heat exchanger, as seen in Fig. 5.9, which does not allow the combustion gases to mix with the circulating airflow. It does, however, allow for the transfer of heat from the combustion gases to the circulating airstream by means of heating the metal heat exchanger.

If the heat exchanger becomes corroded, cracked, or has holes (making this examination will be covered in the next chapters on summer tune up), carbon monoxide and other dangerous and sometimes lethal combustion products will be allowed to enter the home. If this is the case, the heat exchanger of the forced air heating system must be replaced.

All forced air heating systems must be vented in some manner to allow the combustion gases to be exhausted to the outside. All fuel-burning systems lose some heat through the flue. The more efficient the forced air heating system is, the cooler will be the combustion gases that must be vented. On some of the more modern and highly efficient systems, the flue can be made of PVC pipe and can be vented through the wall of the home, and thus no chimney is required.

The temperature of the combustion gases on older forced air heating systems often exceed 400°F. These hot gases rise through the chimney very quickly, keeping it warm enough to not allow moisture from the combustion gases to condense. Recent federal legislation requires new forced air heating systems to have efficiency ratings of 78 percent or higher, which means that they may produce flue gases with temperatures of 200°F or less. A chimney that worked well with a lower-efficiency forced air heating system may begin to show signs of staining or material breakdown when a high-efficiency forced air heating system is installed as a result of the lower flue gas temperatures. This lower flue gas temperature will allow the combustion gases to condense in the chimney.

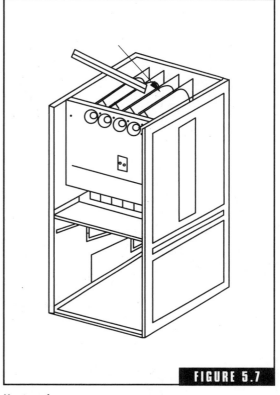

FIGURE 5.7

Heat exchanger.

QUICK»»TIP

If you are going to replace an older, less efficient forced air heating system for a client with one that is rated 78 percent or higher, the chimney must be replaced or upgraded if necessary.

As an example, an induced-draft forced air gas heating system is rated at about 80 percent efficient. An electrically powered fan blows the combustion products through the inside of the heat exchanger and into the flue. Because the gases are cool and do not heat the flue as much, there is a greater chance for condensation to form. Another example is the condensing forced air gas heating system that has an efficiency rating of 90 percent. In this type of forced air heating system, the vented combustion gases are so cool that they condense even before they leave the heating system. These gases are rarely discharged through a chimney. Instead, they are discharged through PVC pipe that is mounted through the wall of the home.

Now that we have learned the controls, electrical system, and operation of the home gas forced air heating system and the different types of gas forced air heating systems, it is time to move on and put this knowledge to use by performing a summer tune-up on a

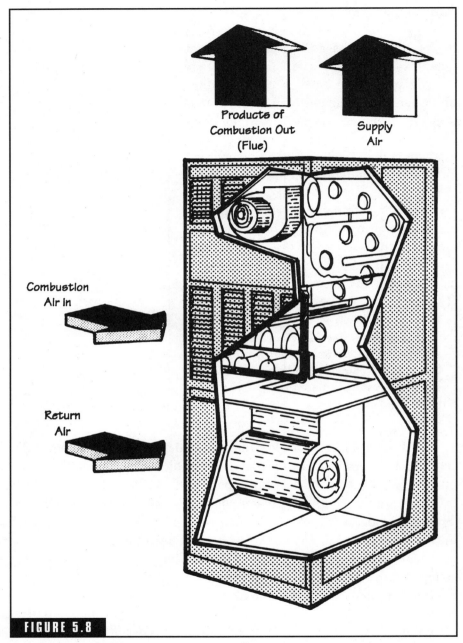

Products of
Combustion Out
(Flue)

Supply
Air

Combustion
Air In

Return
Air

FIGURE 5.8

Combustion airstream.

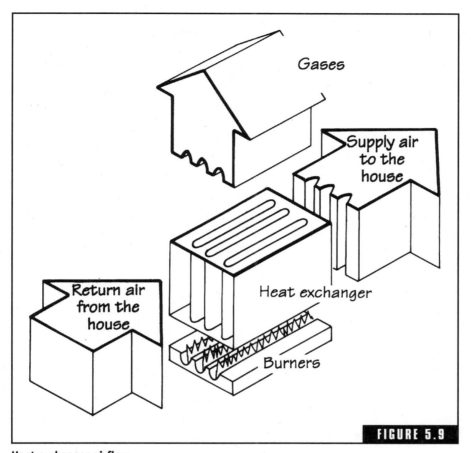

FIGURE 5.9

Heat exchanger airflow.

gas heating system. You should understand all these systems before moving on to the next chapter. If there was something that you did not understand, you should go back over the chapter again before moving on. If you are confident that you have an understanding of the preceding chapters, then let's move on and learn how to properly perform a summer tune-up on this type of system.

Tuning Up a Gas Forced Air Heating System

Now that we have examined the basic components of a home gas forced air heating system, it is time to learn how to perform a basic summer tune-up that could save the homeowner hundreds of dollars in unnecessary repairs that can be avoided by some preventative maintenance. For the homeowner who wishes to have a basic summer tune-up performed, the cost of the supplies needed to do the job should not be more than $30.

The first thing to remember when attempting any repairs on a home gas forced air heating system is *safety*. You will be in close contact with fire, gas, low voltage (24 V), and high voltage (110 V ac).

TOOLS

SUMMER TUNE-UP

1. Flat blade and Philips screw drivers
2. A set of open-end wrenches
3. Oil can with SAE 30 oil
4. Flashlight or drop light
5. Thermostat wrench
6. Allen wrench set
7. Voltmeter
8. Millivolt tester

You must always be aware of these dangers. The most common cause of injury while working on these systems is not using common sense. You should never smoke around a gas heating system at any time. You also must make sure that the power is turned off to this device before attempting any repairs. With this said, let's begin the summer tune-up of your home gas forced air heating system.

The first thing that you must do is locate the power source to the furnace. This typically will be 110 V ac power, the same power that runs to the outlets and lights in the home. There usually will be two separate switches to disconnect the power to the furnace. The first one will be located in the fuse box or breaker box that supplies power to the home (Fig. 6.1). Once you locate this panel, open the front cover and look to see if any of the fuses or breakers is marked for the furnace. If there is one marked for the furnace, either remove the fuse or turn off the breaker. If there are no markings, you will have to locate the fuse or breaker that controls the power for the furnace. To accomplish this, the first thing you must do is have the furnace call for power. This can be done in one of three ways:

1. Locate the fan control, as described in Chap. 5, and see if there is a button on the outside of the cover marked "fan." If there is, pull this button out, and the fan should start to run. Figure 6.2 shows a fan control with this button.

2. If the fan control cannot be used to start the blower manually, on some thermostats there is a lever on the side of the unit that says "fan on or off." Turn this to the on position, and the fan should start to run. Figure 6.3 shows this type of thermostat.

3. Turn the thermostat up until the setting is higher that the temperature in the home so as to start the heating cycle.

Once either the fan or furnace is running, go back to the fuse or breaker panel and start either removing fuses (one at a time) or turning off breakers until the furnace shuts off or the blower stops running. When you remove a fuse or turn off a breaker and the furnace or fan does not stop, replace the fuse or reset the breaker and try the next one. Once you locate the fuse or breaker that controls the power to the furnace, mark this location on the inside of the panel for future use. This

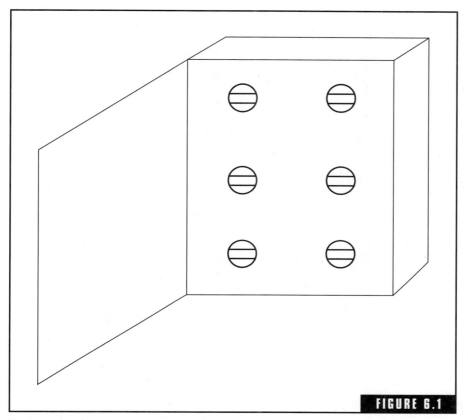

FIGURE 6.1

Fuse panel.

will be a valuable tool in the chapter on troubleshooting the heating system (Chap. 7).

The second switch for the furnace should be located either on or near the unit. It will be a silver box with a switch and fuse or a gray box with a lever that you pull down. In this case, you must pull the lever down before you can open the cover to examine the fuse. In rare cases, there will be no second switch for the furnace, and the only place to turn off the power to the furnace will be at the fuse panel or breaker box. It is very important that you conduct a thorough search for this second switch because it will save valuable time if the homeowner should need to check the fuse on the furnace in the event that there is no heat one day and the cause is a bad fuse. This will be a very

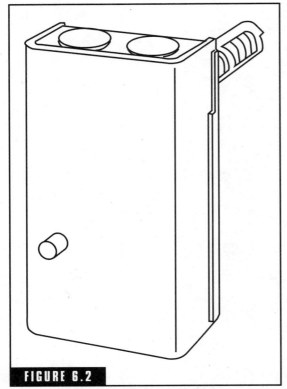

FIGURE 6.2

Fan control with manual fan switch. (*Courtesy of Honeywell, Inc.*)

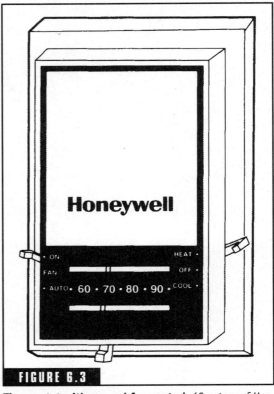

FIGURE 6.3

Thermostat with manual fan control. (*Courtesy of Honeywell, Inc.*)

expensive fuse if the homeowner needs to call a service professional to service the furnace in the night only to find out that there was a second fuse that he or she did not see. One woman who called me for service late one night found this out the hard way, since that fuse cost her over $200. She said that she was going to have it bronzed!

Now that the power is disconnected, be sure to either push the button back in on the fan control, turn the thermostat down, or return the blower switch on the thermostat to the auto position at this time.

The Pilot Assembly

You are now ready to begin the summer tune-up on the gas furnace. It does not matter whether you have natural gas or propane, the steps will be the same. We will begin with what I believe to be the most important safety device on the home gas furnace—the pilot assembly. Please read this section carefully and completely before conducting the safety check on the pilot assembly!

From the preceding chapter, locate the pilot assembly on the furnace. It will be located with the burner(s). It may be behind a cover or panel, so you will either have to remove this panel or open the access panel. Figure 6.4 gives an example of an electronic pilot assembly. You will notice that the assembly consists of a pilot light (flame) and a thermocouple. This thermocouple is the safety device that sends a signal to the gas valve to either continue supplying gas to the pilot because it can sense that the pilot is lit by means of the heat or to shut off the supply of gas to the pilot because no heat is sensed. Without the pilot light being lit, the gas valve will not open so as to allow dangerous gas to fill the home. This is why I feel that the thermocouple is the most important safety device on the furnace. As you can see, if the thermocouple is not in proper operating condition, it will not allow the furnace to operate, and you will have no heat.

As I mentioned earlier, the thermocouple operates by sensing the heat from the pilot. This heat is converted into millivolts (a small electric current). It takes a minimum of 20 mV to keep the pilot section of the gas valve open to supply gas to the pilot light. As I said, this is the minimum amount needed. The pilot *may* still operate at less that 20 mV, but this would be a borderline situation, and the chance for a failure of the thermocouple in the winter would be high.

If you own a millivolt tester, you can test the thermocouple by performing the following steps. (*Note:* If you have a pair of alligator-type clips for your tester, this would be a big help. Otherwise, this step could be tricky.)

QUICK»TIP

Make it a point to *always* check the operation of the thermocouple when performing a summer tune-up.

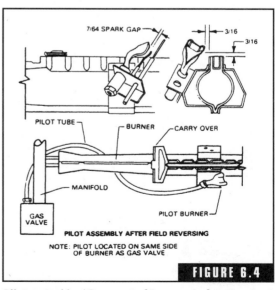

PILOT ASSEMBLY AFTER FIELD REVERSING

NOTE: PILOT LOCATED ON SAME SIDE OF BURNER AS GAS VALVE

FIGURE 6.4

Pilot assembly. (*Courtesy of Lennox Industries, Inc.*)

1. Turn the knob on the gas valve to the "pilot" position. Press down and hold the button. If you release this knob, the pilot will go out, and you will have to relight it. On an older-style furnace, where the pilot line is connected directly to the main gas line, disregard this step.

2. Disconnect the thermocouple from the gas valve. (You do not have to worry about gas escaping from the gas valve because the thermocouple is not connected to the gas.) The thermocouple line is the gold-colored tube. On some older-style heating systems, the thermocouple will be connected to a box with a red reset button on it.

3. Set your meter to millivolts. Make sure that it will read a minimum of 30 mV.

4. Connect the negative lead to a ground (gas line, bare metal, etc.).

5. Connect the positive lead to the end of the thermocouple lead that you removed from the gas valve (Fig. 6.5). Provided that the pilot is still lit and that you have a good ground, you should be getting a reading. If the reading is above 22 mV, you have a good thermocouple reading. If the reading is below 22 mV, you should replace the thermocouple as described later in this section.

If you do not own a meter, there is the manual approach. While this approach is not as accurate as the meter test, the results will be the same if the test is done properly. To check your thermocouple manually, perform the following steps:

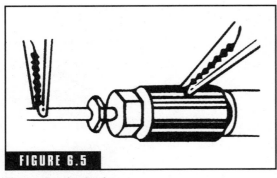

FIGURE 6.5

Thermocouple check.

1. Remove the access panel or lift the access door that covers the pilot assembly.

2. Leaving the knob *on the top of the gas valve* in the "on" position (the position it should be in now), blow out the pilot light. You will have to listen closely to hear the gas running to the pilot.

You are also listening to hear the gas valve close. If the gas valve closes in 0 to 15 s, you will need to replace the thermocouple.

If the gas valve closes in 15 to 25 s, you have a marginal thermocouple. If this step takes 25 to 60 s, you have a good thermocouple.

Most people think that you would want the gas valve to close very quickly to show a good thermocouple, but this is not the case. If it takes a relatively long time for the gas valve to close, this indicates that the thermocouple is getting plenty of heat from the pilot light. This is also a good indication that the pilot light is burning properly, with a good flame.

Before we discuss the steps to replace the thermocouple, let's take a moment to discuss the pilot light. In the test that we just performed, if the gas valve dropped out in a short amount of time, the reason might be traced back to a poor pilot light. How does the pilot light look? The flame should be dark blue in color. A poor pilot flame causes many cases of failure. As you will recall from our previous discussion on the operation of the pilot assembly, without proper heat from the pilot light, the thermocouple cannot keep the gas valve open because not enough current is generated by the pilot. Figure 6.6 illustrates what the pilot light should look like.

The reason for this discussion is that if you are going to remove the pilot assembly to replace the thermocouple, you might as well clean the pilot at the same time. Many cases for failure can be traced to the pilot not being cleaned at the time a thermocouple was replaced. This would be like replacing the engine in your car when the true cause of the problem was a faulty transmission. It will not operate properly.

To remove the pilot assembly, perform the following steps. Figure 6.7 shows the pilot assembly attached to the gas valve and burners.

1. Turn the knob on the top of the gas valve to "off." If the pilot line is not connected to the gas valve, turn the valve off. In either case, the pilot light should go out. If not, check to see where the pilot gas supply is coming from, and shut that off. (*Warning: Do not perform the next step until the pilot light is out.*

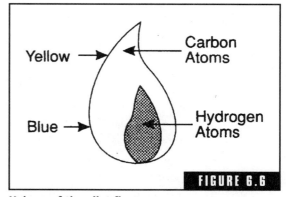

Makeup of the pilot flame.

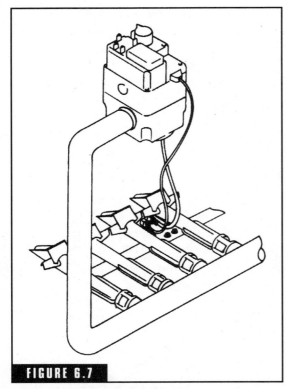

FIGURE 6.7

Burner assembly with pilot assembly.

This will cause gas to escape and could potentially cause an explosion!)

2. Disconnect the pilot line from the gas valve or the valve to which it is connected.

3. Disconnect the thermocouple from the gas valve or control device.

4. Before disconnecting the pilot assembly, take a look at how it is connected to the burner. Is the bracket in front of the burner or behind it? Taking a few seconds to make this determination now will save you time when you replace it. Also, take a look at the way the pilot gas line runs. Since this is soft aluminum tubing, it will bend very easily. It is fine to adjust this tubing, but care must be taken not to crimp it. If the tubing becomes crimped, you will have to replace it.

5. Now remove the screws that hold the pilot assembly in place, and remove the pilot assembly from the furnace. This may be a little tricky, since you will have to maneuver it under the burner and gas manifold. Take your time, and do not force it out.

Now that you have the pilot assembly in hand, you will need to determine the length of the thermocouple lead. The most common lengths are 24 and 36 in. It is always better to get a little longer lead than one that is too short.

Remove the thermocouple by either unscrewing it from the pilot assembly or pulling it from the pilot assembly if there is no nut to remove. When you purchase a new thermocouple, it should come with a universal mount kit so that it will work in either case. To replace the thermocouple, find the items in the kit that will be needed to adapt the thermocouple to the pilot assembly as it was when you removed it. If you are planning to clean the pilot light at the same time (this is recommended as long as you have the assembly out), do not replace the thermocouple until after you put the pilot section back together. You will need the extra room for access to the nuts that hold the pilot gas supply line and orifice to the pilot assembly. If you are not planning to replace the thermocouple because you have determined that the

If you had to unscrew the thermocouple from the pilot assembly, use the adapter with threads attached to it. If the thermocouple was pulled out, use the adapter for the bayonet-type assembly.

cause of the failure is a dirty pilot, you will need to remove the thermocouple to allow enough room to remove the lines and orifice to the pilot. You should consider replacing the thermocouple at this time, however, since it is inexpensive and could save you time later.

TO CLEAN A PILOT LIGHT:

TROUBLE-SHOOTING

1. Remove the thermocouple.
2. Place one wrench on the orifice (this is located closest to the pilot). Place one wrench on the nut that screws the pilot gas line into the orifice.
3. Turn the nut to remove the gas line from the orifice.
4. Hold the pilot assembly with vice-grip pliers while removing the orifice from the pilot assembly. (*Caution: The orifice is made of very soft brass. Care must be taken not to crush the end of this orifice or the pilot gas line will not screw back into place. If this happens, you must replace the orifice.*)
5. Blow through the orifice to remove any dirt from the holes. Check that they are clear by looking through the orifice while looking at a light source. If the hole or holes look clear, the orifice is clean. (*Warning: Do not attempt to clean the holes by placing anything in them. This will only enlarge the holes and make the pilot unsafe to operate. These are precision-sized holes and should never be altered.*)
6. Holding the pilot assembly, blow through, from the bottom, to remove any dust or dirt from the pilot section. You have now cleaned the pilot assembly.
7. Assemble the pilot assembly by reversing the preceding steps. When finished, you should have the thermocouple securely in place and the orifice and pilot gas line in place.

Before attaching the pilot assembly to the burner, you should take the time now to see if there is any debris on the burners themselves. This will act as an insulator and will reduce the efficiency of your heating system. If there is debris on the burners (e.g., rust flakes, soot, etc.), you will want to remove the burners to properly clean them. To clean the burners, perform the following steps. Figure 6.8 shows how to disassemble the burners.

1. Turn off the main gas supply to the gas valve. This valve should be located on the main gas line leading to the gas valve.

SIMPLIFIED BURNER REMOVAL:

1. **Remove cover by loosening bottom screws and removing cover front screws.**
2. **Remove pilot tube, spark wire and sensor wire. Remove gas valve and manifold assembly.**
3. **Remove burner assembly.**

2. Using a pipe wrench or slip-joint pliers, loosen the union fitting on the main gas line close to the gas valve.

3. Loosen or remove the screws that hold the gas manifold to the sheet metal by the burners.

4. Lift the manifold up to release it from the screws that you just loosened. In some cases, you may need to remove part or all of the screws to free the manifold.

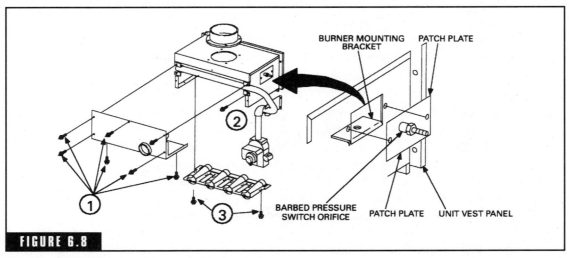

FIGURE 6.8

Burner removal. (*Courtesy of Lennox Industries, Inc.*)

You should now be able to remove the burners from the heat exchanger. It is important to remove and clean only one burner at a time because they each have their own location within the heat exchanger. Please note the silver metal ring at the end of the burner where the orifice from the manifold is located. This ring controls the amount of air that is allowed into the burner during the operation. Care should be taken not to move the location of these rings, since they have been set at this point. When you begin the heating cycle after the summer tune-up is complete, you will examine the way the flame of each burner looks, and you may need to adjust these registers at that time.

Each burner should be cleaned with either a brush or a vacuum cleaner to remove any debris and dirt that is on them. Once you have cleaned the burner, you need to blow air through the burner to remove any dust that is inside them. Simply blow into the round end of the burner to blow the dust out. When no more dust blows out of the burner, replace it in the location from which it came, making sure that the tab at the rear of the burner is in contact with the slot in the heat exchanger. You can check this by twisting the burner once it is in place. If you are able to twist the burner, it is not in the proper location. Adjust the burner as needed to lock it into place. Now repeat the operation for the remaining burners.

Once all the burners have been cleaned and installed, lift the manifold back into position, making sure that the orifice on the manifold is inside the hole in the end of the burner. Replace or tighten the screws that hold the manifold in place, and examine the location of the orifice to the burner one more time to make sure that each orifice is resting inside the hole in the end of the burner. Connect the union back into place and tighten. Whenever you remove any main gas lines, you must check them for leaks before starting the heating system.

QUICK»»TIP

SAFE WAYS TO CHECK FOR LEAKS IN MAIN GAS LINES

1. Use a *gas sniffer* to determine if there is any gas leaking from the connection. If a leak is found, tighten the connection until no more gas is detected. This is a very easy way to check for leaks.
2. Make a thick mixture of detergent and water. Simply brush the solution onto the connection. If there is a leak, the leak will cause the soap to form bubbles. If a leak if found, wipe the solution from the connection, tighten it, and test again. Repeat this operation until the leak is fixed.

QUICK TIP **Use a magnetic screwdriver to hold the screw while you position the pilot assembly in place.**

Now attach the pilot assembly to the burner with the screws that you removed. Next, install the pilot gas line in the proper location on the gas valve or line, and then attach the thermocouple lead in the proper place.

Turn the knob on top of the gas valve to "pilot." You will need a long match or lighter to light the pilot light. When you have one of these items ready, hold it next to the pilot assembly, and press the button down to light the pilot light. Continue to hold the button down for 30 to 60 s. Release the button and turn the knob to "on." If the pilot light does not remain on after 60 s, check all your connections; look at the pilot flame to make sure that it is clean and that it is in good contact with the thermocouple. If not, turn the gas valve off and repeat the preceding steps until the pilot remains on. If everything looks good and the pilot still will not remain on, remove the thermocouple from the gas valve and look to see if there is anything inside the hole that you screw the thermocouple into. I have seen a situation in which the end of the thermocouple that was replaced broke off and remained inside. This will cause the new thermocouple to not make contact with the valve, and this will keep the gas valve from operating.

If you still cannot keep the pilot lit, you may have a problem with the gas valve, and it will have to be replaced. If the thermocouple is connected to a box with a reset button on it, you will have to replace this item.

In some units that have not been serviced in some time, this is quite common. The pilot remains on all the time, and a problem is not detected until you begin your work. You may have a hard time explaining to the homeowner why you are replacing an expensive gas valve when everything was working fine before you arrived. My answer to this is: "Mrs. Jones, it is good that we found this problem before the heating season starts, because now you will not have to worry this winter."

If everything is working properly, you can now move on to the next part of the heating system that must be serviced—the blower. This is the same procedure that you will have to go through if you have a failure in the thermocouple in the winter, and as you have seen, it is much more enjoyable to do this in the summer than when your client discovers that he or she has no heat in the winter.

The Blower Assembly

Now that you have serviced the heating section of the furnace, let's turn our attention to the blower section. As I mentioned at the beginning of this book, there two basic types of blower units: (1) direct drive, where the blower cage is attached *directly* to the motor shaft, and (2) belt drive, where the blower cage is connected to the motor by a *belt and pulley system.* Both these types of systems perform the same function—to blow warm air into your home during the heating cycle.

Locate the blower section of the heating system. If you have determined that you are dealing with an upflow system, the blower will be located in the compartment *below* the burner unit. If you have a downflow system, the blower will be located in the compartment *above* the burner unit. If you have a "low boy" system, the blower will be located in the *rear* of the system. The first thing that must be done before beginning to service the blower unit is to make sure the power is disconnected to the heating system. Once the power has been disconnected, remove the door(s) to gain access to the blower assembly.

Direct-Drive Unit

If you are dealing with a direct-drive unit, look at the end of the motor that is protruding out from the blower housing (Fig. 6.9). Now look at the top of this motor and see if there is a small hole at that location. If there is, then you will want to add 2 to 3 drops of oil in this hole to lubricate the bearings. In some cases, there also may be a hole on the opposite end of the motor that is inside the blower cage. If there is, you will need to lubricate this area as well. If you cannot locate any small oil ports on your motor, then you have a *sealed motor unit, and no lubrication is necessary.*

You should now take the time to examine the blower (squirrel cage) as well. If there is a buildup of dirt on the fins of the cage, you need to clean them out. A

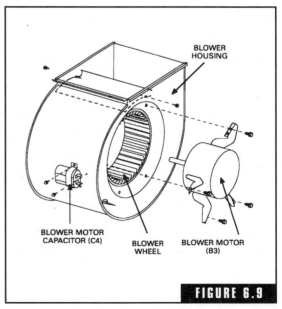

FIGURE 6.9

Direct-drive blower motor. (*Courtesy of Lennox Industries, Inc.*)

buildup of dirt on the fins will cause a reduction in the amount of air-flow that the blower can provide. You can use a screwdriver to scrape the dirt off the fins starting from the rear and pulling it toward you. Remove this dirt with either a vacuum cleaner or some device so that the fins are clean. You also will want to make sure the blower cabinet is free from dirt and dust. It will not do much good to clean the blower fins and leave dirt in the blower cabinet that will be pulled into the blower the next time the blower is started.

The next item to examine is the filter. In the downflow type of heating system, the most common type of filter is the mat filter. In this type of filter, a metal cage holds the filter material. This material typically comes in a roll that is cut to fit the cage. Some heating companies have this material cut to size and prepackaged for the homeowner who wants to replace his or her own filters. The heating professional, though, probably will have a roll of this material on his or her truck.

Remove the filter cage from the blower cabinet. Slide the bar at each end of the filter cage, and separate the two sides. Remove the old filter material and measure the length so that you have the proper size.

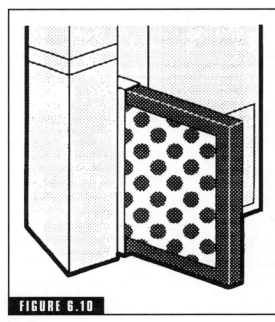

Select the width of filter material that you will need, and cut the new filter to size. You will note that the filter material has a light color on one side and a darker color on the other side. You also should note that one side of the filter material is coated with an oily substance. This is the side that *faces up into the cold air duct.* This is the side of the material that will collect the dust and dirt that is pulled into the blower cabinet.

Once you have the filter material cut to size, lay it on the cage and assemble the cage unit. Now install this cage back into the blower cabinet. Replace the door(s) on the blower cabinet.

The other type of filter is the box filter. Figure 6.10 shows this type of filter. The box filter may be located in a slot in the cold air return or in the top of the blower

FIGURE 6.10

Box filter.

compartment. It is made up of a cardboard frame and filter material. On one side of the filter material there will be wire mesh. This is there to protect the blower from the filter material being pulled into the blower. This is the side that faces the blower. There is an arrow on the side of the filter that shows the direction of the airflow. The filter should be installed with the arrow pointing toward the blower. There are several styles of this filter on the market. Some of these filters do not have wire mesh covering the filter material. You do not want to use this type of filter because there is a chance that the filter material can be pulled into the blower. This will cause damage to the blower and could mean that you will have to replace the motor before it is time.

This unit is now serviced and is ready for use.

Belt-Drive Units

If you have a belt-drive unit, this will require slightly more maintenance than a direct-drive unit. Most older heating systems use this type of system (Fig. 6.11). On an upflow unit, the blower cabinet will be located under the burner assembly. On a "low boy" type of heating system, the blower cabinet will be located in the rear of the unit.

First, locate the motor unit. This typically will be located on top and to the rear of the blower unit. You will find an oil hole located at each end of the motor. You will need to add 2 or 3 drops of oil into each of these holes. In some cases, where the motor has been replaced with a more modern motor, there will be no oil ports. In this case, you have a *sealed bearing unit, and no lubrication is required.* Next, remove the belt from the pulleys. To do this, pull the belt over one of the pulleys, and rotate the pulley clockwise to remove the belt. Care must be taken not to get your fingers caught between the belt and the pulley. Turn the belt inside out, and examine it. If there are signs of splitting or cracking, the belt will need to be replaced. Turn the belt right-side out, and look for the size of the

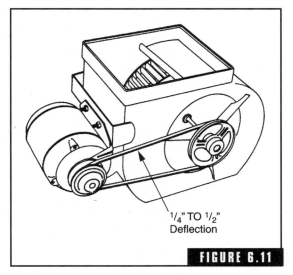

1/4" TO 1/2" Deflection

FIGURE 6.11

Belt-drive motor.

belt. This will be printed on the outer casing of the belt. If you cannot read this information, take the belt to an auto parts or hardware store, and they should be able to measure this for you so that you get the proper size replacement belt. If you are at a service call location where it is not convenient to leave to find an auto parts or hardware store, find a belt in your service truck that is slightly *smaller* than the one that you removed. It can be assumed that the belt has been on the unit for some time and that it has stretched. It is more important to get the proper width of belt to the pulley that is on the unit, since you can adjust for length if it is required. You also should look around the heating unit to see if the last person to service the unit left an old belt there. This will give you an idea as to the proper size as well.

Next, examine the blower unit. In some cases, oil ports will be located on the bearings that the blower shaft goes through. If such ports exist, add oil to them. If there are no oil ports, check to see if your blower has grease cups. These would be located on the upper portion of the bearings. If there are grease cups, remove the cup(s) to see if there is enough grease in them. Most of these types of lubricating systems require a special grease that you may need to purchase from your local heating company. The heating professional should always have at least one tube of this grease on the service truck. Fill the cups, and replace them on the bearings. Turn them down until you feel a resistance (it becomes harder to turn them). The homeowner should turn these grease cups one-quarter to one-half turn every other month during the heating season to allow for proper lubrication to the bearings. Failure to do this may cause the bearings to go dry. This will cause a metal-to-metal situation that will wear out the blower shaft.

If the heating system has not been serviced for a long period of time, this is one item that you should check as part of the tune-up process. You should ask the homeowner if the blower is noisy when the unit is running. If he or she says "yes," this will give you a good indication if there is excessive wear on the blower shaft.

You can check for this condition in one of two ways:

1. With the belt removed, pull up and down on the blower shaft pulley. If there is excessive play (a good bearing and shaft should not move at all), you have a worn-out bearing.

2. Replace the belt on the motor and blower, and turn the fan switch to "manual" to start the blower. Listen to the sound of the blower. If you hear a squealing noise or a rumble, you have a bad wear situation. You also can look at the way the pulley turns. If you detect a wobble in the pulley, you have wear.

You need to inform the homeowner of this finding before you proceed any farther. You must get the homeowner's permission to make this expensive repair. In most cases, this will be a time and material job, and depending on how far you are from your shop (you will need to disassemble the unit and go to the shop to repair it), you could be looking at 2 to 3 hours of labor. Explain to the homeowner that the situation will only get worse and that he or she could be looking at this unit failing in the winter, and then there will be no heat. Also explain that this is the reason you come to the home in the summer to tune up the heating system so that the chances of having a failure in the winter are reduced.

Once you have the homeowner's permission to do the repair, you will need to disassemble the unit and remove it from the blower cabinet.

Once you get the blower removed from the cabinet, examine the blower fins for signs of dirt buildup. If there is a buildup on the fins, prior to taking the unit to the shop, stop at a local manual car wash and spray down the blower unit. The high-pressure spray works great to remove the old buildup on these types of units. Where the buildup is very heavy, some of these car washes have a setting for degreasers, and this works well.

Once this is complete and the blower unit is at the shop and on the bench, you will need to remove the locking collars that hold the shaft in place. Normally, you

> **QUICK»TIP**
>
> **DISASSEMBLING THE UNIT FROM THE BLOWER CABINET**
>
> 1. Check to make sure that the power to the unit is turned off.
> 2. Remove the locking straps that hold the motor to the blower unit.
> 3. Remove the motor from the motor bracket, and place it on the bottom of the blower cabinet. Make sure that you do not allow the motor to hang from the wires or conduit that is connected to the motor. Use some kind of brace if needed to support the motor.
> 4. Remove the bolts that hold the blower unit to the heating unit.
> 5. Remove the blower unit from the blower cabinet.

will find two of these collars located on the opposite end of the blower from the pulley. One should be on the outside and one on the inside of the rear bearing. On some units, these devices will be incorporated into the blower cage, one at either end. In this case, simply loosen the setscrews, but do not remove them. Loosen and remove the outside collar, and loosen the inside collar. This will then allow the blower shaft to be removed. If the bearings have been neglected for a long period of time, the shaft may be "frozen" to the bearings and will not slide out of the unit. In this case, lubricate the entire shaft with motor oil. You will then need to use something to drive the shaft out of the bearings.

Never strike the end of the shaft with a hammer or other driving device. This may cause the end of the shaft to flare. The best approach is to use a piece of shaft material that is either the same size or slightly smaller to drive the shaft out of the bearings. Once you have the shaft free from the rear bearing, you can attempt to pull it out the remainder of the way by twisting the shaft and turning the pulley back and forth while pulling at the same time. If this works, pull the shaft free from the blower unit. If the shaft still will not come free, you will need to continue with the driving method until the shaft is free from the blower unit.

Once the shaft is free, you will need to remove the pulley from the shaft, since you will need to use this pulley on the new shaft. Loosen the setscrew that holds the pulley on the shaft, and remove it from the shaft. If the pulley will not come off, place motor oil between the end of the shaft and the pulley, and use a wheel puller to remove the pulley. Place the jaws of the puller around the pulley, and screw the jacking bolt down to make contact with the end of the shaft. Tighten the jacking bolt with a wrench until the pulley comes free. Care must be taken if the pulley is aluminum so that you do not bend it. If it does bend, you will need to replace the pulley as well.

Next, remove the bearings from the blower unit, and find replacements that are the same number. It does not matter if the replacement bearings have grease fittings or oil fittings; they will both work equally well. I prefer grease fittings over oil fittings because it is much easier to get the homeowner to turn the grease fittings down than to have him or her put oil in the oil fittings.

Once you have the new bearings installed, you will need to make a new shaft. You will need to take three measurements of the shaft:

1. Measure the length of the shaft.

2. Measure the diameter of the shaft.

3. Measure the length of the flat spot on the shaft that is used to tighten the pulley.

Once you have all three of these measurements, cut the shaft to length, and grind the flat spot on the shaft. Lightly oil the shaft, and slide it into the first bearing and through the blower cage, and insert the first locking collar. Next, slide the end of the shaft through the rear bearing, and attach the other locking collar. Tighten the locking collars only finger tight, since you will have to make the final adjustments once the blower is located back in the blower cabinet. In the case of set screws that are part of the blower cage, use this same procedure, and only tighten them finger tight as well. Slide the pulley onto the shaft, making sure that the set screw is lined up with the flat spot on the shaft. Tighten the set screw finger tight as well.

Once the blower is reinstalled in the blower cabinet, replace the motor on the motor bracket, and secure in place. You will need to measure the distance from the outside of the blower housing to the inside of the motor pulley. This measurement will need to be the same as the distance from the blower housing to the inside of the blower pulley. This distance is critical so that the belt will travel in a straight line from the blower motor to the blower itself. If the belt does not travel in a straight line, the belt can jump off either pulley or cause excessive wear in the new bearings that you just installed.

Once you have the alignment correct, tighten all set screws on the blower shaft and the pulley. Also double-check to make sure that the mounting bolts for the blower unit and motor mounts are tight. Now check the belt tension, and make sure that you have the proper $\frac{1}{4}$- to $\frac{1}{2}$-in deflection. Make the necessary adjustments to the motor adjustment jacking screw to achieve this deflection.

Turn on the power to the heating unit, and manually start the blower only by means of the blower switch or by adjusting the fan control to start the blower. Watch how the belt turns on the blower pulley. If there does not appear to be a straight line of travel between the motor pulley and the blower pulley, turn off the power and adjust as necessary; then recheck. When the alignment is correct, this repair is complete.

You will now want to examine and/or replace the filter. The filter will be located either inside the blower compartment or inside the *cold air duct* leading to the blower compartment. Filters come in several types depending on the type of heating system that you have. It will either be a box filter (see Fig. 6.10), in which the filter material is enclosed in a cardboard box and has a mesh cover on the side that faces the blower, or a mat filter, in which the material will be enclosed in a metal frame. Filters come in several different sizes as well.

Locate the filter in the heating unit and remove it. If you have a box filter, the size of the filter will be imprinted on the outside of the filter. Replace this filter with the same size filter. Make sure that you look for the arrow on the side of the filter that will show you which way the filter is to be installed. Typically, the way to remember this is that the side with the metal mesh that covers the filter material will face the blower compartment. This is designed so that the filter material will not be sucked into the blower during the heating cycle. If you have a mat filter that is mounted in a removable frame, you will need to measure the amount of filter material you will need. This can be done by measuring the existing material that is mounted in the frame. Write this information down for future use. Cut the proper size filter from the roll, and install it in the frame, making sure to put the coated side out. Place the other side of the frame on the filter, and lock it into place. Reinstall the filter frame into the blower cabinet.

The summer tune-up of the gas forced air heating system is now complete. All that remains is to replace any doors and access panels that you removed during this operation and turn the power back on by resetting all breakers or replacing the fuse(s). The final step is to check the operation of the heating unit by running it through a complete heating cycle and checking the safeties.

Before you turn on the power to the gas heating unit, you will need to disconnect the power to the blower unit. The reason that this is done is so that you can check for the proper operation of the *high limit safety* (Fig. 6.12). This is the high limit that is set on the fan control and will shut off the power to the gas valve and blower if the blower does not come on during the heating cycle.

If you have a direct-drive unit, open the access panel to the wiring cabinet (make sure that the power it turned off first), and locate the black (hot) wires. Look at the wires, and you should find a wire marked for the blower. Remove the wire nut that holds all the wires together, and remove the wire for the blower. Replace the wire nut on the remaining wires, and make sure that the black wire for the blower is not touching any metal objects. You also must make sure that you do not touch this wire while the power is turned on.

If you have a belt-drive blower, you will not have to disconnect any wires. Simply remove the belt to simulate this same condition.

Before you turn the power back on to the unit, you need to check the thermostat to make sure that it is operating properly. Remove the

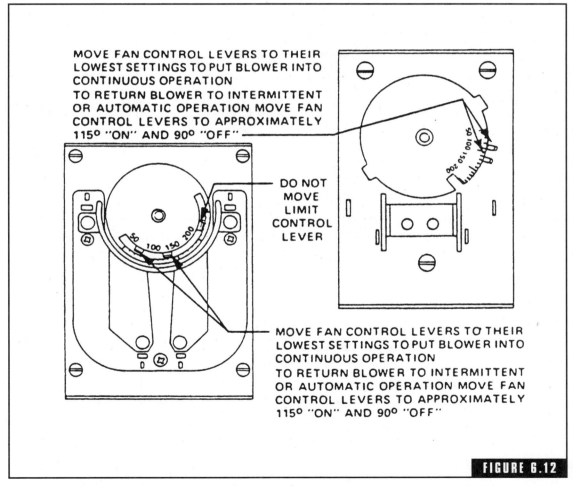

Limit switch. (*Courtesy of Lennox Industries, Inc.*)

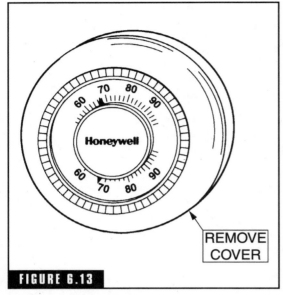

REMOVE COVER

FIGURE 6.13

Thermostat. (*Courtesy of Honeywell, Inc.*)

front cover from the thermostat, and check for any lint or dirt, removing any if necessary (Fig. 6.13). Next, turn the thermostat up *slowly* until either you see the mercury bulb drop to the right to indicate that the thermostat is calling for heat or you hear the bimetallic points come together indicating the same thing. Now look at the pointer on the top of the thermostat that shows where the reading is at the point that the heating cycle should start. Now compare this to the reading on the lower part of the thermostat (the thermometer reading that shows the actual temperature in the home). These readings should be the same. If they are not, you need to adjust the thermostat so that they are the same.

If you have a thermostat wrench, place it on the nut behind the coil that controls either the mercury bulb or the points. Determine from the readings if you needed to turn the coil to the left (thermostat set point higher than the actual temperature) or to the right (thermostat set point lower than the actual temperature). Make very slight movements with the wrench. After you have made the adjustment, turn the thermostat all the way down (to the left) and then up (to the right) until the thermostat calls for the heating cycle to begin, and compare the readings. Make the necessary adjustments again until it is correct. Once this is done, leave the thermostat turned up (calling for heat), and return to the heating unit.

Turn the breaker on or replace the fuse so that there is power to the heating unit. The burners should light at this point. Once the heat has reached the set point on the high limit, the heating unit should shut down. If it does not, you will need to replace the fan control, since the safety feature on this device is not working properly. This could cause a major overheating situation that is unsafe.

To replace the fan control, shut the power off to the heating unit, and remove the wires to the fan control. Write down the location of these wires if needed for reference when you install the new control.

Fan controls come in two basic types: The first type has a round metal tube with a metal coil inside that contracts when heated, causing the dial on the front of the fan control to turn until the set point is reached, sending power to the blower motor causing it to run (Fig. 6.14). The second type has a bimetallic strip that comes together when heated, sending power to the blower motor causing it to run. The front of this type of fan control has a sliding lever to adjust the set points (Fig. 6.15).

Both of these fan controls work the same way, and they come in different lengths. It is important that you do not put in a new fan control that is a different length from the one that you are replacing. These controls come in different lengths depending on the size of the heating unit they are used in. If you use one that is too long, it may make contact with a metal surface inside the heat exchanger and give a false reading, therefore not operating properly. One that is too short will not operate properly because it will not be able to reach far enough into the heating stream to be effective.

Remove the screws that hold the fan control to the heating unit, and remove the control. Once you have determined the proper size

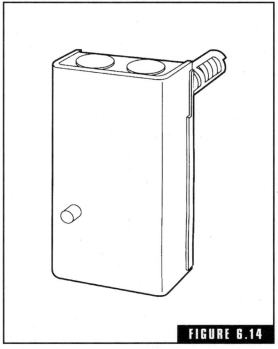

FIGURE 6.14

Fan/limit control. (*Courtesy of Honeywell, Inc.*)

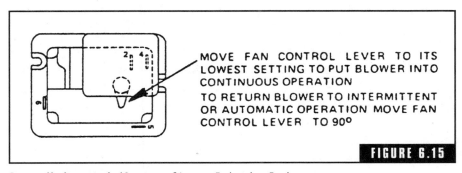

MOVE FAN CONTROL LEVER TO ITS LOWEST SETTING TO PUT BLOWER INTO CONTINUOUS OPERATION

TO RETURN BLOWER TO INTERMITTENT OR AUTOMATIC OPERATION MOVE FAN CONTROL LEVER TO 90°

FIGURE 6.15

Square limit control. (*Courtesy of Lennox Industries, Inc.*)

fan control to use, install the new fan control and reattach the wires. Adjust the off set point to between 70 and 90°F and the on set point to between 140 and 150°F.

Turn the power to the heating unit back on, and check the high limit again. The heating unit should shut down when the fan control reaches this number. If this fails again, recheck your steps, and try again. You may not have connected the wires properly, or you may have used the wrong fan control for this type of unit.

Once this check is satisfactory, reconnect the black wire for the blower (make sure that the power is turned off) or replace the belt. Now start the heating cycle again. This time you are checking to make sure that the heating system operates properly for a complete heating cycle.

If you replaced the fan control, you are looking for one more item. You want to see that when the blower is running in a normal heating cycle it remains on during the complete cycle. If the blower starts, runs for a period of time, and then shuts off while the burners are still on, you will need to adjust the fan on temperature higher to allow more heat to build up prior to the blower starting. If the blower shuts off at the end of the heating cycle and then comes back on again, you will need to adjust the off temperature lower to allow for the blower to run longer so that there is no more heat buildup in the heat exchanger prior to the blower shutting off.

Once you have checked the safeties and are sure that everything is working properly, you will want to check the temperature rise of the unit. Each heating unit has a temperature range that is located on the rating plate. To check this range, you will need to place one thermometer in the warm air (supply side) and one in the cold air (return side) ductwork (Fig. 6.16). You must make sure that the thermometer that is placed in the warm air duct cannot "see" the heat exchanger, thus picking up the radiant heat.

Set the thermostat to the highest setting, and place the thermometers in position (you will have to make small holes in the ductwork if they are not already there). Once the thermometers have reached their highest steady temperature, subtract the two readings to get the rise. If this reading is within the range on the rating plate, this step is complete. If the range is too high, you will need to speed up the blower. If it is too low, you will need to decrease the blower speed. On direct-

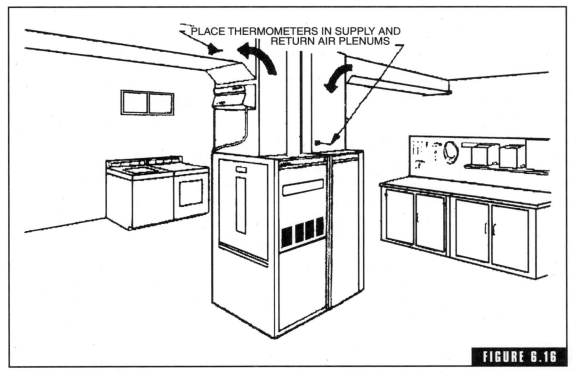

PLACE THERMOMETERS IN SUPPLY AND RETURN AIR PLENUMS

FIGURE 6.16

Thermometer placement for temperature rise. (*Courtesy of Lennox Industries, Inc.*)

drive units, you will need to select the proper color wire to make the adjustment. On belt-drive units, you will need to loosen the set screw on the end of the pulley. Turn the pulley clockwise to decrease the speed and counterclockwise to increase the speed. Tighten the set screw. After you have made the adjustment, check the readings again, and adjust the speed as necessary to get the proper reading.

There is one more critical check that must be made before the summer tune-up is complete, and this is the heat exchanger check. The heat exchanger (Fig. 6.17) needs to be checked to make sure that there are no holes in it. The heat exchanger's job is to convert the heat generated from the burners into heat for the home while diverting harmful gases (carbon monoxide) created from the combustion process through a system of baffle plates, out the chimney, and away from the home. If the heat exchanger becomes old and worn, holes can develop that will allow this gas to escape into the home. Carbon

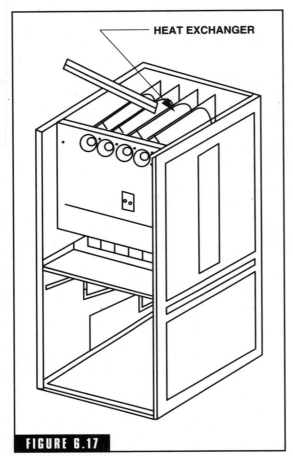

HEAT EXCHANGER

FIGURE 6.17

Heat exchanger. (*Courtesy of Lennox Industries, Inc.*)

monoxide is a colorless, odorless, tasteless gas, so it is almost impossible for the homeowner to detect until it is too late. No summer tune-up or, for that matter, emergency heating call should be considered complete until you are absolutely sure that there is no problem with the heat exchanger.

There are two ways to check for proper operation of the heat exchanger:

1. Observe the burners on the heating system when the blower starts. If you notice that the flames react in a different way or become "lazy" and roll out the front of the heating system, you have a problem with the heat exchanger.

2. The other way is to make an inspection hole in the hot air plenum that is connected to the heating section of the unit and, using an inspection mirror, look inside during the heating cycle when the burners are on. If you can see light coming from anywhere in the heat exchanger, then there is a danger.

In either one of these cases, you *must* inform the homeowner of your findings and red tag the unit. Turn off the gas supply, and inform the company that you work for. In some states, you may have to inform the gas company as well so that they can lock out the gas service to the unit.

In either case, the homeowner has one of two choices to make:

1. Replace the heat exchanger.

2. Replace the heating system.

This is a choice that the homeowner must make. You cannot allow a heating system with a faulty heat exchanger to continue to operate, since this can cause death if all the conditions are right.

It is always a good idea to suggest to the homeowner of a gas heating system to purchase and install a carbon monoxide detector in the home as a safety precaution. The cost is inexpensive, and it may help to save a life.

If everything checked out as described and you did not discover any other problems, the summer tune-up of the heating unit is complete. If you do not want to wait for the thermostat to turn the burner unit off, you may turn it down at this time and allow the blower unit to complete the cycle. If the heating unit did not function as described earlier (the burner did not light, the blower did not come on, etc.), go back over the steps again to see if you missed anything. As I mentioned earlier, you may uncover a more serious problem with the heating system during this process, but it is better to find out now than when you have to use the heating system for the first time during cold weather and it does not work.

If you have completed all the steps described in this chapter and the heating system is operating properly, you are finished. You should have acquired the confidence at this point to move on to the next chapter on troubleshooting a gas forced air heating system in the event of failure during the heating season.

Troubleshooting a Gas Forced Air Heating System

As a heating professional, you will be called on to perform any number of system checks. None will test your skills more or be more challenging than troubleshooting calls. Most problems with gas forced air heating systems can be traced back to the items that we covered in Chap. 6. As you will recall, a gas forced air heating system is comprised of two parts:

1. The heating section
2. The blower section

These two sections of the heating system work together to heat the home; however, they are typically not connected with regard to system failures. For example, a problem with the burners not igniting will have nothing to do with the blower section. Therefore, you can isolate the problem to the heating section and concentrate your efforts on the components that operate this section of the heating system.

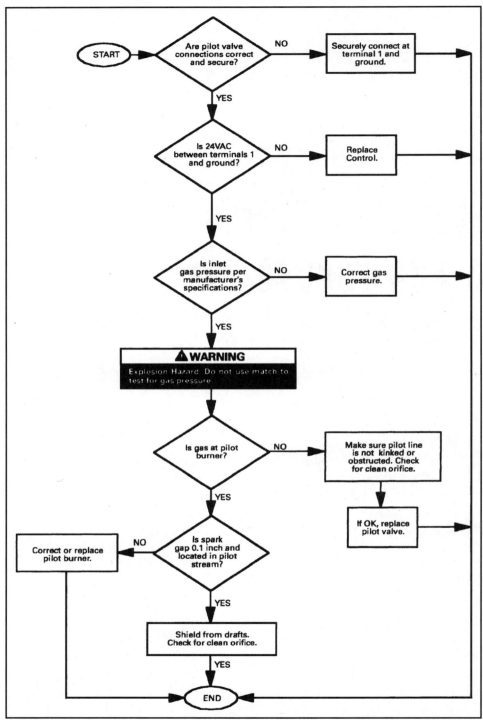

TROUBLESHOOTING CHART 7.1 Spark ignition. Ingition Control. Spark is present but pilot will not light. (*Courtesy of Lennox Industries, Inc. and © 1993 Johnson Controls, Inc. Reprinted with permission.*)

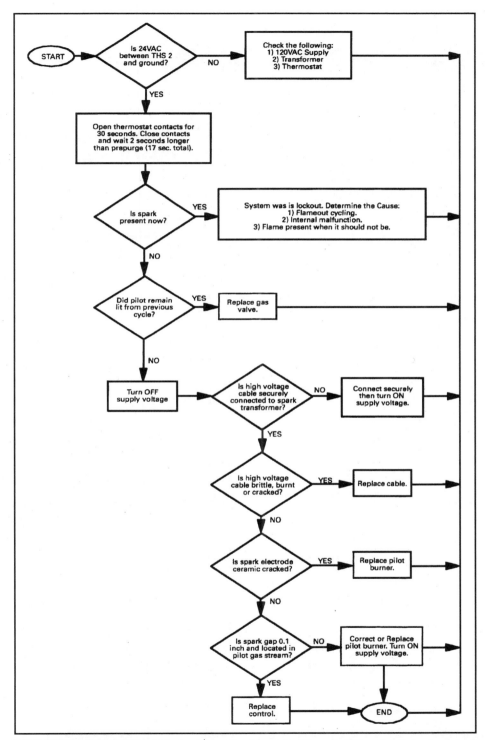

TROUBLESHOOTING CHART 7.1 (*continued*)   Spark ignition. Ingition Control −1 and −2
Models. No spark and system will not work. (*Courtesy of Lennox Industries, Inc. and ©
1993 Johnson Controls, Inc. Reprinted with permission.*)

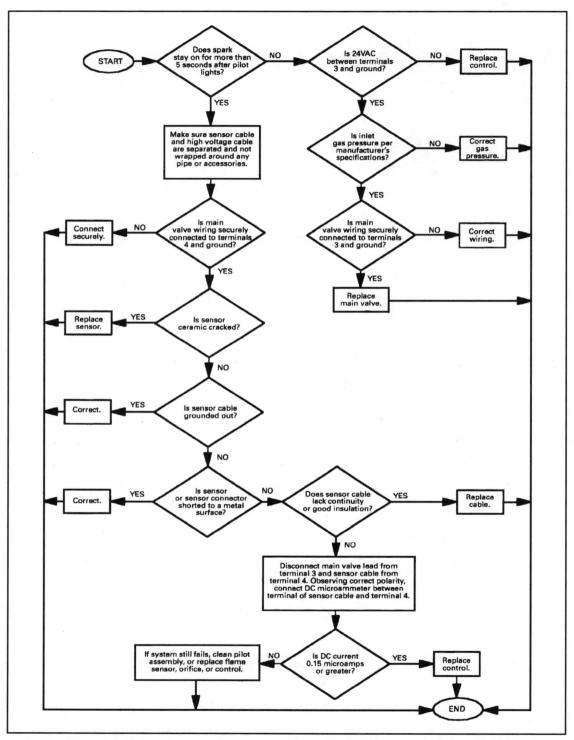

TROUBLESHOOTING CHART 7.1 (*continued*) Spark ignition. Ingition Control −1 and −2 Models. Pilot lights but main valve will not come on. (*Courtesy of Lennox Industries, Inc. and © 1993 Johnson Controls, Inc. Reprinted with permission.*)

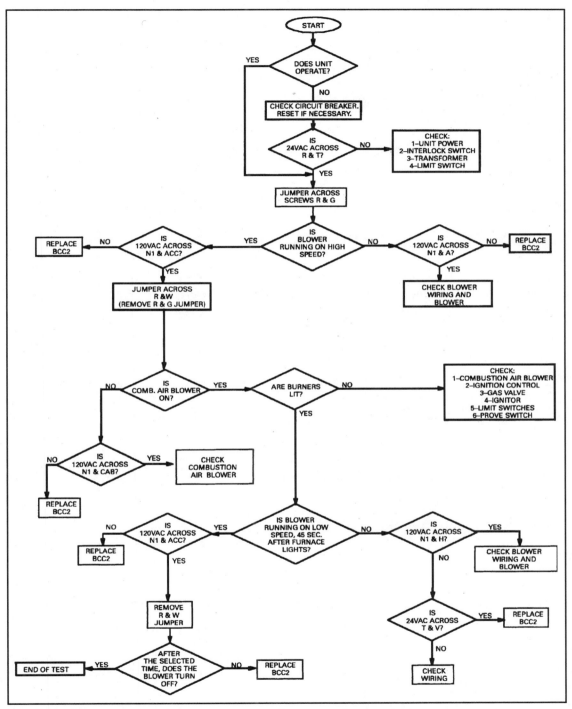

TROUBLESHOOTING CHART 7.2 **Blower.** (*Courtesy of Lennox Industries, Inc.*)

At the end of this chapter, I have included flowcharts that will illustrate the points in this chapter. I will be detailing these steps for both the standing pilot system and the electronic ignition system. You will discover that both these systems operate in the same manner, with some minor differences with regard to the way that the pilot and burners ignite.

It is very important to remember the message of Chap. 2 on listening and observing. As discussed there, the homeowner is your best source of information on the problem that you have been called on to troubleshoot and repair. Listen to what the homeowner is saying, ask the questions that were outlined in Chap. 2, and this will lead you to the potential problem. Then observe what the system is or is not doing to come up with an answer. This is the most challenging and exciting part of the job. Here you will continue to build on the knowledge that you have already gained from the previous chapters and will start to see how the different parts of the heating system, when not operating properly, will cause a failure of part or all of the system. It is your job as a heating professional to be able to quickly diagnose the problem and come up with the proper solution. Some solutions will be very simple and may not require you to replace any parts. Others will require you to replace the defective component. By following the recommendations in Chap. 6 on the types of items needed on your service truck, you will show the client that you came prepared for the challenge.

Where a problem is common to either a standing pilot system or an electronic system, I will give the steps for troubleshooting the standing pilot system first and then the electronic system. There will be areas that are very similar, so I will cover them only once.

With all of this said, let's begin this chapter on troubleshooting a gas forced air heating system.

Pilot Lights, But Main Valve Will Not Open (Standing Pilot System)

The first item that a heating professional should check in this situation is the thermostat. Is the thermostat turned up and calling for heat? If not, turn the thermostat all of the way up. Do the burners light? If they do, then turn the thermostat back so that it is no longer calling for heat. Now turn the thermostat up to just past the temperature in the home.

Do the burners light? If they do, allow the heating system to make a complete cycle to make sure that it is operating properly. It would be a good idea to go to the heating system and do a visual check to make sure that all items appear to be in good working order. If they do, then you have solved the problem. While this will be a rare case, it can and does happen. If, however, after turning up the thermostat, the burners still do not light, check to see if the thermostat has switches for heat on-off and blower on-off. If it does, make sure that the switch is in the proper position to call for heat. If not, set the switch for heat. Do the burners light? If the answer is yes, you have corrected the problem. If not, set the blower control to on or manual. Is the blower running? If the answer is yes, you have determined that you have power to the unit and have eliminated this possible cause. If the blower did not come on, check for a blown fuse or breaker that has tripped.

Turn the power off to the unit at the switch on the heating system, and replace the fuse or reset the breaker. Remember that there may be more that one fuse to the unit, so make sure that you do a complete check to look for all fuses.

If you find a blown fuse or tripped breaker, you will need to discover the reason for the problem. In some very rare cases, you will find a blown fuse caused by the fuse being weak. You always should replace a fuse with one of the same rating (15 to 20 A), and it should be a fusetron type fuse. This is a slow blow type of fuse that will allow a delay in blowing to compensate for the extra amperes needed for a blower to start, since this is when a properly operating system will draw the most amperes.

Turn the power back on to the unit, and observe the operation. If everything operates properly, you have repaired the problem. One possible cause for a blown fuse is a bad blower motor. This will be covered later in this chapter, but for now, let's assume that the burner still will not light.

Turn the power off to the unit. Remove the door to the burner section of the unit. Remove or lift up the cover that protects the burners and pilot assembly if the unit is so equipped. Observe the condition of the pilot light. On a properly operating pilot, the flame will be strong and blue in color. If the pilot light is yellow in color, it is not putting out enough energy to allow the gas valve to open. You will need to remove and clean the pilot at this point.

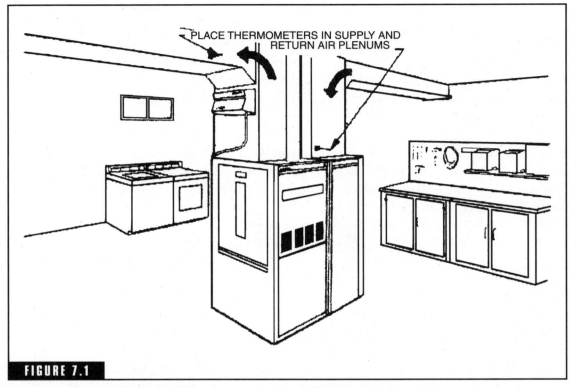

PLACE THERMOMETERS IN SUPPLY AND RETURN AIR PLENUMS

FIGURE 7.1

Thermometer placement for temperature rise. (*Courtesy of Lennox Industries, Inc.*)

Turn the knob on the top of the gas valve to off. Disconnect the pilot line from the gas valve. Disconnect the thermocouple line from the gas valve. Remove the screws that hold the pilot assembly to the burner. Remove the pilot assembly from the unit. Disconnect the pilot line from the pilot assembly. Remove the orifice from the pilot assembly (you may need to remove the thermocouple from the pilot assembly to have enough room to work on the pilot line removal). Blow air through the orifice to remove any dirt that is in the orifice. Blow air through the pilot housing to remove any lint or dirt that may be in this area. Reassemble the pilot assembly, and install it in the unit. Make sure that the assembly is installed in the same position from which it was removed. Figure 7.2 shows the relationships of a properly installed pilot assembly.

Check the connections of the pilot gas line and thermocouple line to the gas valve. Turn the knob on top of the gas valve to the "pilot" position, press down, and hold. Light the pilot light. Continue to hold the knob down for between 45 and 60 s. Release the knob. Does the pilot remain lit? If not, repeat the procedure for lighting the pilot again. Does the pilot stay lit? If the answer is yes, turn the knob to the "on" position, and turn the power back on to the unit. Do the burners light? If they do, the problem was a dirty pilot light. If the burners do not light, you will need to check the thermocouple.

Turn the power off to the unit. Turn the knob on the gas valve to "pilot," and hold it down. Remove the thermocouple lead from the gas valve. Set your meter to millivolts, and connect the negative lead to a ground and the positive lead to the flat end of the thermocouple (see Fig. 6.5). Is the reading below 20 mV? If the answer is yes, replace the thermocouple. To do this, repeat the procedure for removing the pilot assembly, and replace the thermocouple. Check the reading again. Is the reading above 20 mV? If it is, connect the thermocouple to the gas valve, and check the operation. Do the burners light? If the answer is yes, you have repaired the problem. If the burners still do not light, check the voltage to the thermostat.

Locate the transformer that supplies power to the thermostat. Typically it will be located in the wiring cabinet, or it could be mounted on the outside of the unit. Figure 7.3 shows a typical 24-V ac transformer. Check for output power on the transformer. Do you get a 24-V reading? If the answer is no, check to make sure that all wiring connections are tight. If they are and you still do not get a reading, you will need to replace the transformer. To do this, remove the wires connected to the terminals, and then remove the screws that hold the transformer in place. Disconnect the wires from the power supply, and remove the old transformer. Reverse this procedure to attach the new transformer in

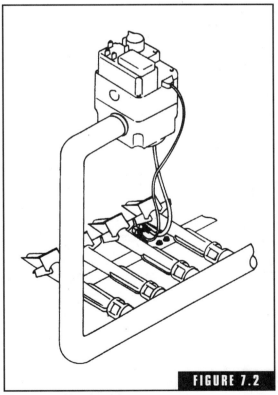

FIGURE 7.2

Pilot assembly placement.

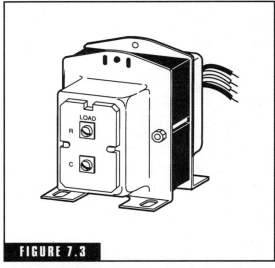

FIGURE 7.3

24-V ac transformer. (*Courtesy of Honeywell, Inc.*)

TROUBLE-SHOOTING IF THE FOLLOWING CONDITIONS EXIST AND THE BURNERS STILL WILL NOT LIGHT, YOU HAVE ISOLATED THE PROBLEM TO A BAD GAS VALVE.

1. Power is on.
2. Pilot flame is good.
3. Thermocouple reading is at least 20 mV.
4. Gas valve knob is in the "on" position.
5. Thermostat is turned up, calling for heat.
6. There is 24 V to the thermostat.
7. All fuses and breakers are good.

place. Attach the wires to the terminals, and turn on the power to the heating system. Do the burners light? If the answer is yes, you have repaired the problem.

Before you replace the gas valve, which is a very expensive component, double-check to make sure that everything on the list is in proper operating condition. If everything on the troubleshooting list is working properly, do one final check of the wiring to make sure that all connections are tight. If they are and the burners still will not come on, then you must replace the gas valve (see Fig. 7.4).

Figure 7.5 shows one way the gas valve is connected to the gas line. First, turn off the power to the unit. Next, turn off the gas supply to the unit. There will be a shutoff valve located on the large gas line leading to the gas valve. Once this is done, you will need to disconnect the pilot gas line

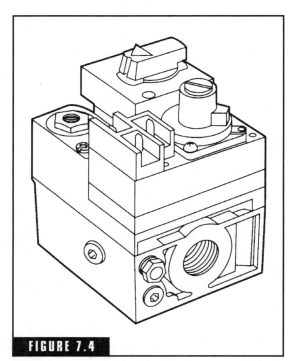

FIGURE 7.4

Gas valve. (*Courtesy of Honeywell, Inc.*)

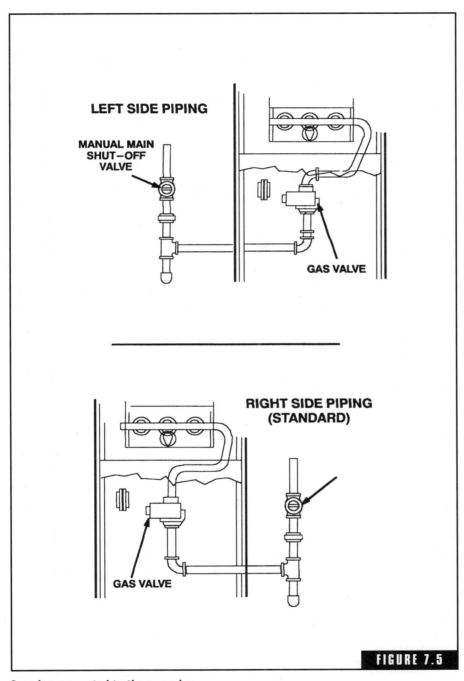

LEFT SIDE PIPING

MANUAL MAIN SHUT-OFF VALVE

GAS VALVE

RIGHT SIDE PIPING (STANDARD)

GAS VALVE

FIGURE 7.5

Gas pipe connected to the gas valve.

QUICK›››TIP   Use a piece of masking tape to write the location of the screw that the wire is connected to and then attach it to the wire or draw a small wire diagram to show the location.

and the thermocouple line from the gas valve. Before removing the wires from the gas valve, note their location. You should replace the valve with the same brand where possible. This makes the installation process much simpler.

Next, disconnect the main gas line. There should be a union connection located close to the valve. Use a pipe wrench to loosen this connection. Then unscrew the fitting to separate the line leading to the valve from the main gas line. You will need to hold the gas valve with another pipe wrench while you remove the gas line from the valve. Turn the gas line counterclockwise to remove it from the valve. Next, turn the gas valve counterclockwise to remove it from the gas manifold leading to the burners.

The new gas valve will have ¾-in openings at each end. If your gas line or manifold is smaller than ¾ in, you will need to use the reducing bushings that come with the new gas valve. You will need to apply a coating of "pipe dope" to these fittings before you screw them into the gas valve. This is used to help seal the metal-to-metal connections so that gas will not leak out around the threads. Screw these fittings into the gas valve as far as you can by hand. Apply a coating of this same pipe dope to the threads of the gas pipe as well. Attach the short piece of pipe to the gas valve first that will attach to the union. You will have hold the gas valve with a wrench while you tighten this pipe. Now attach the valve to the manifold pipe in the same manner using pipe dope on the threads. Turn the valve onto the manifold until it is tight. Make sure that the top of the valve is pointing up as much as possible. Do not overtighten. Screw the two parts of the union fitting back together. Attach the pilot line and thermocouple. Reattach the wires in the same location from which they were removed.

Turn the main gas valve back on, and check for leaks. This can be done with either a gas sniffer or a 4:1 ratio of liquid soap and water applied to all the joints. If any bubbles appear, you have a leak, and the connection must be tightened. Repeat this process until there are no more leaks.

SAFETY›››TIP   Do not use a match or lighter to check for gas leaks because this could cause injury or explosion.

Turn the knob to "pilot," and light the pilot light. Turn the knob to "on," and turn the power on to the unit. Run the heating system through a complete cycle to check its operation.

Electronic Ignition

When you are working on new electronic ignition systems, some units have a computer control that will give you the cause of the problem in a series of lights. Do not disconnect the power to the unit until you have removed the cover to the diagnostic unit. Turning off the power will cause the unit to loose this function, and you will not have the proper information to troubleshoot the unit.

The first item to check is the spark. Does this spark stay on for more than 5 s after the pilot is lit? If the spark does stay on for more than 5 s after the pilot lights, check for 24 V between terminal 3 and ground. If not, replace the control. If there was a 24-V power supply, check the main gas pressure. To check the pressure, turn off the power to the unit, and then turn off the main gas supply and remove the pressure tap plug from the gas valve. Attach a manometer to the tap. Remove the gas valve sensing hose, and plug the end. Turn on the main gas supply. Turn on the power, and start the unit. Allow the unit to run for 5 minutes to stabilize before taking the reading. Operate the unit only long enough to get an accurate reading. Figure 7.6 shows the proper readings for natural and LP gas. Read the gauge, and check and compare this reading with the specifications. If the reading is too low, increase the pressure. To do this, remove the cap on the gas valve regulator, and turn the screw clockwise to increase the line pressure until the gauge reads the correct pressure. Turn off the main gas supply and power to the unit. Remove the gauge, and replace the plug. Replace the cap on the

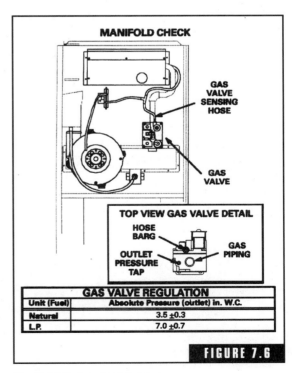

GAS VALVE REGULATION	
Unit (Fuel)	Absolute Pressure (outlet) in. W.C.
Natural	3.5 ±0.3
L.P.	7.0 ±0.7

FIGURE 7.6

Manifold test and pressure.

regulator, remove the obstruction in the sensor hose, and reattach it. Turn the power back on to the unit. Does the valve open? If yes, this is complete; if not, check the wires to the gas valve to make sure that they are tight. If they are, you will need to replace the gas valve. Follow the procedure for gas valve replacement in the preceding section on standing pilots.

If the spark stays on for more than 5 s after the pilot lights, check to see if the high voltage cable is wrapped around a pipe or is in contact with any accessories. If it is, move or route the cable so that is free from any obstructions.

If this was not the problem, check the ceramic on the spark igniter. Is it in good shape, or is there a crack in the ceramic? If there are any cracks in the ceramic, replace the sensor. If the sensor is good, check the connections on terminal 4 and ground. Are they tight? If not, tighten them. If all the connections are tight, is the sensor cable grounding out? Check to make sure that it is not in contact with any metal objects.

Check for continuity in the sensor cable. Also check the insulation on the cable. Is it worn or in good condition. If the cable is worn, replace it. If the cable is in good shape and you still cannot get the burner to light, you will need to check the flame signal. To do this, disconnect the main valve lead from terminal 3 and the sensor lead from terminal 4. Make sure to observe the proper polarity, and check the current level between the sensor cable and the "sense" terminal. Figure 7.7 shows this connection. Set your meter to dc volts. Remember that 1 dc volt = 1 dc microamp. Turn the thermostat up to call for heat. When the unit lights, read the display on the meter. Is this reading greater than 0.15 μA? If the answer is yes, replace the control.

If everything checks out, or after replacing the defective components, and you still cannot get the burners to light, remove and clean the pilot assembly. If the pilot flame looks good, replace the flame sensor and check again. If this still does not solve the problem, you will have to replace the ignition control box.

FLAME SIGNAL MICROAMPS		
G26 –1 and –2 units	Normal	0.25
	Minimum	0.15
G26 –3 units	Normal	> 0.7
	Low	≤ 0.7
	Minimum	0.15

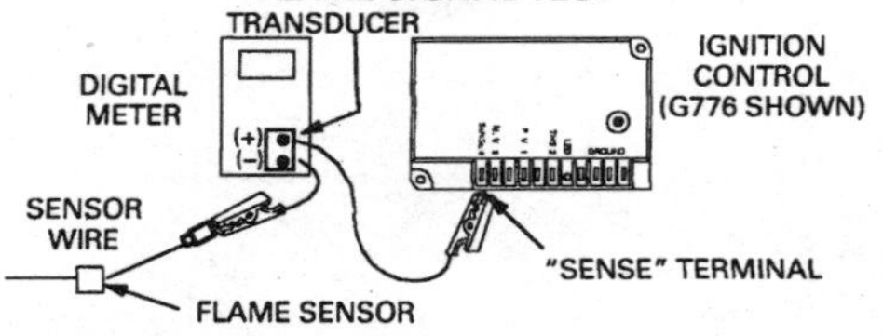

FIGURE 7.7

Flame sensor test. (*Courtesy of Lennox Industries, Inc.*)

There Is No Spark or Pilot Light, and the System Will Not Work

This is one of the most common calls that a heating professional will receive. This is better known as the "no heat" call. The homeowner calls to tell you that the system will not operate and that there is no heat.

If you discover that this is an older standing pilot system, turn the gas valve knob to the "pilot" position, and attempt to light the pilot. When it lights, continue to hold the knob down or the pilot may go out. Check the condition of the flame. Is it a blue color, or does the pilot have a yellow color? If the pilot flame is yellow in color, check the reading of the thermocouple. If the reading is less than 20 mV, you will need to remove the pilot and clean it. It is always a good practice to replace the thermocouple at the same time that you remove and clean the pilot, since the cost is low and you can save yourself and the homeowner money by not having to come back to replace the thermocouple at a later date (see Chap. 6 for details on cleaning the pilot).

If the pilot light is blue, then it is burning properly, and you will need to check the thermocouple. To do this, remove the thermocouple lead from the gas valve while holding the knob down. Set your meter to dc millivolts, and attach the negative lead of the meter to ground and the positive lead to the end of the thermocouple lead. Is the reading above 20 mV? If it is, you have a good reading thermocouple, and you will have to check further for the problem.

Check for 24 V at the transformer. If there are no volts, check the connections. If they are good, replace the transformer. If the pilot flame is good, the thermocouple reading is good, and you have a good transformer, you will have to replace the gas valve. As stated in the preceding section, check all the electrical connections on the gas valve to make sure that they are tight before you replace the gas valve because it is a very expensive part to replace.

Electronic Ignition

If the unit has electronic ignition, check to make sure that you have power to the unit. Do this by checking the fuses or breakers to make sure that they are in good working order. You also can check this by turning the blower on, by setting the blower switch to "manual" on the thermostat, if so equipped, or by setting the fan control to the "on" position. If the blower starts, you know that you have power to the unit.

Once the blower starts, return the switch to the "automatic" position. Now check to see that you have 24 V between terminal THS2 and ground. If there is no power at this point, check the transformer to make sure that it is working properly. If you do not have 24 V at the transformer, replace the transformer, and restart the system.

If the transformer is good, turn the thermostat all the way down for 30 s, and then turn it up until the contacts close. Wait for the prepurge cycle to complete. If a spark is present now, the system has been locked out, and there is a problem with one of the systems. If the system lights, cycle the unit several times to try to determine the cause of the lockout. It could be caused by the flame sensor. If it is, replace this sensor. If the flame sensor was not the cause, check to see if the pilot remained on after the lockout. If it did, you will need to replace the gas valve.

If this was not the case, turn off the power to the unit, and check the high-voltage cable to the spark transformer. Is this cable connected properly? If not, secure it and turn the power back on to check the operation.

If the connections are fine, check the condition of the high-voltage cable. Is the cable brittle or cracked? If it is, replace it, and turn on the power to cycle the unit.

If the cable is fine, check the condition of the ceramic cover on the spark electrode. If there is a crack in the ceramic, you are losing energy from this circuit, and it will not operate properly. If you notice any cracks, replace the spark electrode, and cycle the unit to check for proper operation.

If the electrode is fine, check the spark gap to make sure that it measures 0.1 in and is located in the pilot gas stream. A poor spark gap will not give off the amount of spark that is required to ignite the pilot. This will cause the pilot to fail and the system to lock out. If the gap is not correct, either correct the problem or replace the pilot burner, and cycle the unit to check for proper operation.

If all the other checks are done and no problems are found, you will need to replace the ignition control unit to solve this problem. Replace the unit, turn on the power, and cycle the unit to check for proper operation.

Spark Is Present, But the Pilot Will Not Light (Electronic Ignition Only)

If you have a situation where the system begins to cycle, but the pilot will not light, first check the connections on the ignition control box. Are all the connections tight? If not, tighten them, and reset the thermostat to cycle the unit. If this does not solve the problem, or if all the connections are tight, check for 24 V between terminal 1 and ground on the ignition control. If there is no voltage, replace the control.

If there is voltage at this point, check the inlet gas pressure to make sure that it is at the manufacturer's specification. As described in the preceding section, turn the power off to the unit, remove the outlet pressure tap, and install the manometer. Disconnect the sensing hose, and plug the end. Turn the unit on, and allow it to stabilize for 5 minutes. Take a reading from the gauge, and compare it with Fig. 7.6. Adjust the pressure as necessary to meet the manufacturer's specification. When finished, turn the unit off, remove the manometer, and replace the plug. Remove the plug from the end of the sensor hose, and connect the hose back onto the gas valve. Does the pilot light? If it does, you have corrected the problem.

If the pilot does not light, or if the pressure is within specifications, check to see if there is gas coming from the pilot burner. If there is no gas coming from this point, check to see if the pilot line has been kinked. If it has, replace it. Also check for obstructions in the pilot line. Disconnect the pilot line from the gas valve, and remove the screws that hold the pilot burner in place. Remove the pilot burner and the gas line, taking care not to kink the line. Blow air through the line to clear any obstructions. You also should remove the orifice from the pilot and make sure this is clear as well. Once both are clear, reassemble the pilot burner assembly, and reverse the process to install the assembly back into the unit. Turn the power back on to the unit. Does the pilot light? If the pilot does not light, check the spark gap to make sure that it is 0.1 in and is in the pilot burner gas stream. Correct as necessary. If the pilot still does not light, you will have to replace the gas valve.

I have covered many of the problems that occur with the burner section of a gas forced air heating system for both standing pilot and electronic ignition systems. You will notice that the way I approached all the problems was to concentrate on the simple solutions first and then to move to the more complicated solutions.

Blower Will Not Come On

The first item that you need to look at is the burners. Are the burners lit? If the burners are not lit and the thermostat is turned up and calling for heat, you need to check the fuse panel or breakers. Are the fuses or breakers in good operating condition and set? If they are not, replace the fuse or reset the breaker, and try once more.

On newer units there is an interlock switch on the blower door that must be made for the unit to operate. Make sure that the door is closed properly and making the switch. If it is not, correct this problem, and try once more.

As discussed several times in this book, there are a series of switches that control the heating system. You have already checked the thermostat (a switch), the fuses or breakers (another switch), and now you need to check the switch that controls the blower—the fan control.

Remove the cover to the fan control, if so equipped, and locate the two wires that supply power to the fan control. Figure 3.12 shows these controls. Using your meter set for 120 V ac, check at the terminals for power. If you have power at these terminals and the blower will not come on, you may have a bad fan control. To check this, push in the button on the fan control that says "manual," if so equipped, and see if the blower comes on. If there is no button, move the levers on the fan control to their lowest settings to simulate this same effect. If the blower starts, you will need to replace the fan control.

Turn the power off to the unit. Remove the wires to the fan control. Slide the fan control out of the unit. Measure the length of the contacts on the fan control so that you can replace it with one of the same length. Install the new fan control, and attach to the unit. Replace the wires on the new fan control. Set the on temperature to 115°F, and the off temperature to 90°F. Turn the power back on to the unit, and run it through a complete cycle to check the operation. Adjust the on and off settings as necessary until the blower runs steadily during operation. If the blower comes on and then goes off and comes back on again, you will need to set the on temperature to a higher setting.

Open the door to the blower compartment; this will be located above the burners on an upflow unit and below the burners on a downflow unit. On "low boy" models, the blower will be in the rear of the unit.

Determine if this is a direct-drive unit (motor connected to the blower cage) or a belt-drive unit (blower motor located separate from the blower connected by a belt). If this is a direct-drive unit, do a visual check to see if there are any signs of burned or disconnected wiring. If there are, you will have to replace the blower motor and possibly the wire as well. Before doing this, check the thermostat to see if there is a manual fan setting switch on it. If there is, turn the power off to the unit, and set this switch to the "manual" position. Once you are back at the unit, turn the power back on and observe the blower. Does it start? If not, shut the power off to the unit. You will have to replace the motor. Figure 7.8 shows the procedure for removing the blower on

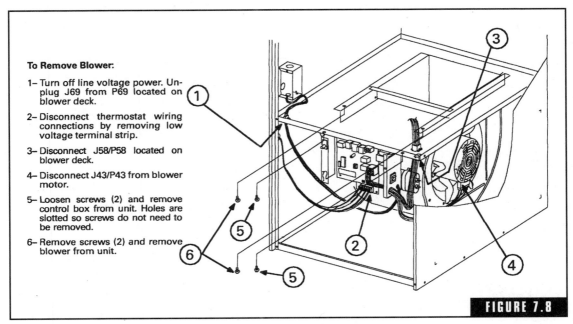

Blower removal. (*Courtesy of Lennox Industries, Inc.*)

one type of heating system. For other types of systems, follow the next procedure.

First, turn the power off to the unit. Remove the wires to the motor. Next, remove all the bolts that hold the blower unit in place. Remove the blower unit from the heating system. Looking at the blower unit from the opposite side of the unit from the blower motor, locate the lock that holds the blower cage to the motor shaft (Fig. 7.9). Loosen this screw enough so that the shaft will slide out from the cage. Remove the bolts that hold the motor to the blower assembly. Remove the motor from the blower assembly. Remove the bracket that is attached to the motor.

Check the plate on the motor for the horsepower and rpm ratings. You need to replace the motor with one with the same ratings. If this unit is used for heating and cooling, be sure that you replace the motor with a variable-speed motor.

Reverse the process, and replace the motor bracket onto the motor. Slide the motor back into the blower assembly and through the hole in the blower cage. Bolt the motor in place. Slide the cage on the shaft until the distances between both ends of the cage are about the same. Hand tighten the cage to the shaft, and rotate the cage. Make sure that the cage does not rub on the sides of the blower assembly. If it does, loosen the cage and move it until there is clearance on both ends and the cage does not rub. Tighten the set screw, and replace the blower assembly into the blower housing in the heating unit. Connect the wires, and start the unit to make sure that everything is operating properly. Check to see if the motor has oil fittings; if so, place a few drops of oil in the oil ports. Run the unit through a complete cycle to make sure that the problem is corrected.

If the unit has a belt-drive blower (Fig. 7.10), check the condition of the belt. If the belt is broken, replace the belt with the same size belt that was on the unit. Check

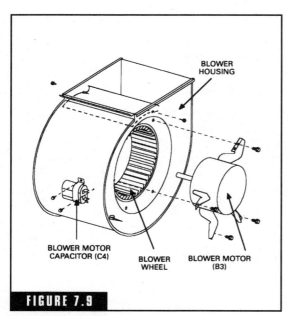

BLOWER HOUSING

BLOWER MOTOR CAPACITOR (C4)

BLOWER WHEEL

BLOWER MOTOR (B3)

FIGURE 7.9

Direct-drive blower removal. (*Courtesy of Lennox Industries, Inc.*)

the belt for size (it will be printed on the outside case of the belt). Once the new belt is in place, make adjustments so that the belt has a ¼ to ½ in of play when you push on the center of the belt. If the deflection is more or less than this, adjust the belt tension adjuster located on the motor bracket until the tension is correct.

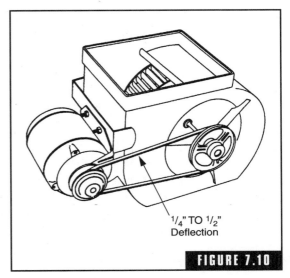

¼" TO ½"
Deflection

FIGURE 7.10

Direct-drive blower.

If the belt is in good shape, check the condition of the motor and wires. Check for loose connections on the wires in the wire cabinet and on the motor. Tighten if needed, and check the operation by setting the fan control to "manual." If the motor runs, you have solved the problem. If the motor still does not come on and you have checked to make sure that you have 120-V power to the unit, you will have to replace the motor.

To replace a belt-drive motor, remove the belt, and then remove the two locking collars that hold the motor to the bracket. There will be one at each end of the motor. Once the locking collars are removed, remove the motor from the bracket. Remove the screws that hold the cover in place to the wires. Use a ¼-in nut driver or pliers to loosen the nuts that hold the wires in place. Remove the nut that holds the conduit in place, and remove the wires from the motor. Next, remove the pulley from the motor shaft. There is an allen screw on the rear of the pulley that holds the pulley to the shaft. Loosen this set screw, and remove the pulley. Check the plate on the motor for the proper horsepower and rpm ratings, and replace the motor with one with the same ratings.

To replace the motor, reverse the procedure. Once the wires are connected, slide the pulley in place on the motor shaft. Do not tighten the set screw at this time. Mount the motor to the bracket.

Replace the belt on both the motor and blower pulleys. Check to make sure that the belt will travel in a straight line between the motor pulley and the blower pulley. Slide the motor pulley in or out as needed to achieve this straight line. Once this is done, tighten the motor pulley. Check the deflection of the belt to make sure that it is between $\frac{1}{4}$ and $\frac{1}{2}$ in. Make adjustments to the belt tension adjuster on the motor bracket as needed to get this deflection. Check the motor to see if there are any oil fittings on the new motor. If there are, place 2 to 3 drops of motor oil in each hole.

Start the unit, and observe the blower. Make sure that the belt is traveling in a straight line and that the tension looks good. Run the unit through a complete cycle to make sure that the blower is not turning too fast. If the blower does not run continually, you will have to slow the blower down. To do this, turn the power off to the unit, and remove the belt from the motor pulley. Loosen the front set screw on the motor pulley, and turn the outer ring of the pulley counterclockwise to decrease the speed of the blower. Tighten the set screw once you feel you have made the proper adjustment, and install the belt. Run the unit again, and check the operation. If the blower does not shut off during the cycle, it is running at the proper speed.

You also can perform a check of the temperature drop to make sure that the blower is running at the optimal speed as well. To do this, set the thermostat at the highest setting. Place a thermometer in the warm air supply in a location where it will not pick up radiant heat from the heat exchanger. Place another thermometer in the return air side. Once the unit has stabilized and the thermometers have reached their highest and steadiest readings, subtract these two readings, and compare them with the rating on the unit rating plate. If the temperature is low, adjust the blower speed up; if the temperature is high, decrease the blower speed. Figure 7.5 shows the location of the thermometers.

Blower Will Not Shut Off

If the blower will not shut off during a heating cycle, this can be traced back to one of the switches not operating properly. First, check the thermostat to see if there is a manual fan control switch on it. If there is, make sure that it is set for "automatic." If the switch is in the "manual" position, the fan will run all the time. Set this switch to "manual." Did the blower stop running? If it did, cycle the heating system to make sure that this was the problem. If this was not the problem or the blower remains running, check the fan control.

If the fan control is equipped with a manual fan switch (Fig. 7.11), make sure that it is off or pulled out in the automatic position. Does the blower continue to run? If the blower stops, you have solved the problem; if not, then tap on the fan control lightly with the handle of a screwdriver. Does this cause the blower to stop? If the answer is yes, you will have to replace the fan control. Since these units have either a bimetallic strip or a heating coil, they can stick in the open position and not allow the fan to stop.

To replace the fan control, turn the power off to the unit. Remove the wires to the fan control. Remove the screws that hold the fan control to the unit, and remove the fan control. Measure the length of the contacts on the rear of the fan control so that you can replace it with one of the same length.

Slide the new control into place, and replace the screws that hold the control

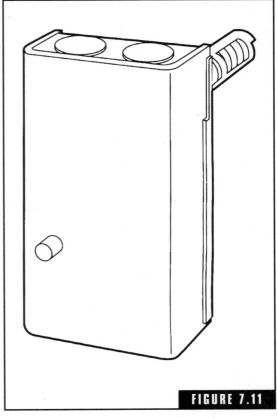

FIGURE 7.11

Fan control. (*Courtesy of Honeywell, Inc.*)

to the unit. Replace the wires. Set the on temperature to 115°F and the off temperature to 90°F. Turn the power back on to the unit, and check the operation. Make sure that the blower does not come on as soon as the power is turned back on to the unit. If it does, you have another problem to deal with. Double-check that any manual fan control switches are set for automatic. If all the fan switches are set to "manual," you have a bad thermostat. This is not a common repair, since thermostats do not go out very often. But it does happen.

It is always best to replace the thermostat with one of like kind and shape. You will have to replace the back plate (Figs. 7.12 and 7.13) to

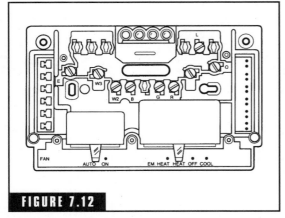

Square thermostat back plate. (*Courtesy of Honeywell, Inc.*)

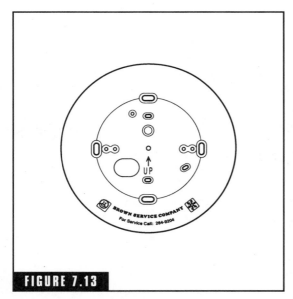

Round thermostat backplate. (*Courtesy of Honeywell, Inc.*)

the thermostat, since this is what controls blower manual and automatic operation. It is also where the wires are connected.

Turn the power to the unit off. Remove the cover to the thermostat. Loosen the screws that hold the thermostat to the back plate, and remove the thermostat. Remove the wires from the back plate. Remove the screws that hold the back plate to the wall. Care must be taken when removing the back plate because the homeowner probably has painted once or several times since the thermostat was installed. You do not want to rip the paint from the wall, so use a knife to "score" the paint around the back plate. Remove the back plate from the wall.

To install the new thermostat, first set the new back plate in place. This is probably the most important step in the installation. You must make sure that the back plate is *level* on the wall or the thermostat will not operate properly. Use a small level to make sure that the plate is installed straight. Replace the wires on the back plate, and then install the thermostat on the back plate. Make sure that all the screws on the thermostat connect with the proper holes on the back plate. Some of these screws are used as connections with the wires on the back plate. Turn the power back on to the unit, and turn the thermostat up to call for heat. Check the operation of the thermostat, and the blower. Once you are sure that the problem is solved, check the settings on the new thermostat to make sure that the temperature at which it calls for heat matches the room temperature. If not, either adjust the back plate or use a thermostat wrench to adjust the nut located behind the coil until the settings are correct. Make small adjustments, and turn the thermostat all the way down and then up until the bulb just drops to the right, completing the circuit and calling for heat. Troubleshooting charts 7.1 through 7.3 are guides for this chapter.

UPON INITIAL POWER UP, REMOVE ALL THERMOSTAT DEMANDS TO THE UNIT

PROBLEM 1: UNIT FAILS TO OPERATE IN THE COOLING, HEATING, OR CONTINUOUS FAN MODE

Condition	Possible Cause	Corrective Action / Comments
1.1 – Both diagnostic lights fail to light up. LED#1–Off LED#2–Off	**1.1.1** Main voltage 120V not supplied to unit.	**ACTION 1** – Check 120V main voltage. Determine cause of main power failure.
	1.1.2 Miswiring of furnace or improper connections.	**ACTION 1** – Check for correct wiring of 120V to power make up box and transformer. **ACTION 2** – Check 24V wiring to control board.
	1.1.3 Circuit breaker tripped or fails to close.	**ACTION 1** – Replace circuit breaker if it is re-set but does not have continuity. **ACTION 2** – If circuit breaker still trips, check for short.
	1.1.4 Door interlock switch failure.	**ACTION 1** – Check that door switch is activated when door is closed. **ACTION 2** – Check wire connections to switch, replace loose connectors. **ACTION 3** – Check continuity of switch in closed position. Replace if malfunctioning
	1.1.5 Transformer Failure.	**ACTION 1** – Check that transformer output is 24V. Replace if malfunctioning
	1.1.6 Failed control board.	**ACTION 1** – If all the above items have been checked, replace board.
1.2 – Diagnostic lights flash the roll-out code.	**1.2.1** Roll–out switch open.	**ACTION 1** – Manually reset the roll-out switch by pushing the top button. **ACTION 2** – Determine the cause of the roll-out switch activation before leaving furnace.
	1.2.2 Roll–out switch failure.	**ACTION 1** – Check continuity across roll-out switch. Replace roll–out switch if switch is reset but does not have continuity.

LED Status	Code / Condition	Action
LED#1–On, LED#2–Slow Flash	**1.2.3** Miswiring or improper connections at roll–out switch.	**ACTION 1** – Check wiring connections to switch.
	1.2.4 Nine pin connector failure	**ACTION 1** – Check 9–pin connector for proper connection to control board. **ACTION 2** – Check continuity of the multi plug pin.
1.3 – On initial power–up the comb. air blower does not energize. – Diagnostic lights flash the reverse polarity code. LED#1–Fast Flash, LED#2–Slow Flash.	**1.3.1** 120V main power polarity reversed.	**ACTION 1** – Check the 120V has line and neutral correctly input into control. **ACTION 2** – Reverse the line and neutral at the 120V field connection.
1.4 – On initial power up the combustion air blower does not energize. – Diagnostic lights flash normal power on operation. LED#1–Slow Flash, LED#2–Slow Flash	**1.4.1** Open combustion air blower motor circuit.	**ACTION 1** – Check for 120V to combustion air blower. If no power, check wire and connections.
	1.4.2 Failed combustion air blower motor.	**ACTION 1** – If power is present at blower, replace blower.

TROUBLESHOOTING CHART 7.3 SureLight Control. (*Courtesy of Lennox Industries, Inc.*)

PROBLEM 1: UNIT FAILS TO OPERATE IN THE COOLING, HEATING, OR CONTINUOUS FAN MODE (CONT.)

Condition	Possible Cause	Corrective Action / Comments
1.5 – On initial power-up the combustion air blower remains energized. – Diagnostic lights flash the improper main ground. LED#1–Alternating Fast Flash LED#2–Alternating Fast Flash	**1.5.1** Improper ground to the unit.	**ACTION 1** – Check that the unit is properly ground. **ACTION 2** – Install a proper main ground to the unit
	1.5.2 6-Pin connector is improperly attached to the circuit board.	**ACTION 1** – Check 6-pin connector for proper installation. Correctly insert connector into control.
	1.5.3 Line voltage is below 75V.	**ACTION 1** – Check that the line voltage is above 75V. Determine cause of voltage drop and supply correct voltage to the control.

PROBLEM 2: UNIT FAILS TO OPERATE IN THE COOLING OR HEATING MODE, BUT COMBUSTION AIR BLOWER OPERATES CONTINUOUS. UNITS WITH CONTROL BOARDS DATE CODED AFTER NOV. 1 1997, WILL OPERATE IN COOLING BUT NOT IN THE HEATING MODE, WITH COMBUSTION AIR BLOWER CYCLING 5 SECONDS ON 55 SECONDS OFF.

Condition	Possible Cause	Corrective Action / Comments
2.1 – On initial power-up the combustion air blower remains energized. – Diagnostic lights flash the improper main ground. – Units with control boards date coded after Nov.1 1997; combustion air blower will cycle 5 seconds on 55 seconds off. LED#1–Alternating Fast Flash LED#2–Alternating Fast Flash	**2.1.1** Open ignitor circuit.	**ACTION 1** – Check for correct wiring and loose connections in the ignitor circuit. Check mult-plug connections for correct installation.
	2.1.2 Broken or failed ignitor.	**ACTION 1** – Unplug ignitor and read resistance across ignitor. If resistance does not read between 10.9 and 19.7 ohms, replace the ignitor.

PROBLEM 3: UNIT FAILS TO FIRE IN THE HEATING MODE, COMBUSTION AIR BLOWER DOES NOT ENERGIZE

Condition	Possible Cause	Corrective Action / Comments
3.1 – Unit operates with a cooling or continuous fan demand. – Combustion air blower will not start with a Heating demand. – Diagnostic lights flash the limit failure mode. LED#1–Slow Flash, LED#2–On	**3.1.1** Primary or secondary (if equipped) limit open.	**ACTION 1** – Check continuity across switch(es). Switches reset automatically upon cool down. **ACTION 2** – Check for restrictions on blower inlet air (including filter) and outlet air. Determine cause for limit activation before placing unit back in operation.
	3.1.2 Miswiring of furnace or improper connections at limit switch(es).	**ACTION 1** – Check for correct wiring and loose connections. Correct wiring and/or replace any loose connections.
3.2 – Unit operates with a cooling and continuous fan demand. – Combustion air blower will not start with a Heating demand. – Diagnostic lights flash the pressure switch failure code. LED#1–Off, LED#2–Slow Flash	**3.2.1** Miswiring of furnace or improper connections to combustion air blower.	**ACTION 1** – Check for correct wiring and loose connections. Correct wiring and/or replace any loose connections.
	3.2.2 Pressure switch stuck closed.	**ACTION 1** – Check that the pressure switch is open without the combustion air blower operating. Replace if malfunctioning

TROUBLESHOOTING CHART 7.3 (*continued*) SureLight Control. (*Courtesy of Lennox Industries, Inc.*)

PROBLEM 3: UNIT FAILS TO FIRE IN THE HEATING MODE, COMBUSTION AIR BLOWER DOES NOT ENERGIZE (CONT.)

Condition	Possible Cause	Corrective Action/Comments
3.3 – Unit operates with a cooling and continuous fan demand. – Combustion air blower will not start with a Heating demand. – Diagnostic lights flash the pressure switch failure code 2.5 minutes after heating demand. LED#1–Off, LED#2–Slow Flash	**3.3.1** Miswiring of furnace or improper connections to combustion air blower.	**ACTION 1** – Check for correct wiring and loose connections. Correct wiring and/or replace any loose connections.
	3.3.2 Combustion air blower failure.	**ACTION 1** – If there is 120V to combustion air blower and it does not operate, replace combustion air blower.

PROBLEM 4: UNIT FAILS TO FIRE IN THE HEATING MODE, COMBUSTION AIR BLOWER ENERGIZES, IGNITER IS NOT ENERGIZED.

Condition	Possible Cause	Corrective Action/Comments
4.1 – Unit operates with a cooling and continuous fan demand. – Combustion air blower energizes with a heating demand. – Diagnostic lights flash the pressure switch failure code 2.5 minutes after heating demand.	**4.1.1** Pressure switch does not close due to incorrect routing of the pressure switch lines.	**ACTION 1** – Check that the pressure switch lines are correctly routed. Correctly route pressure switch lines.
	4.1.2 Pressure switch does not close due to obstructions in the pressure lines.	**ACTION 1** – Remove any obstructions from the the pressure lines and/or taps.

LED#1–Off **LED#2–Slow Flash**	
4.1.3 Pressure switch lines damaged	**ACTION 1** – Check pressure switch lines for leaks. Replace any broken lines.
4.1.4 Condensate in pressure switch line.	**ACTION 1** – Check pressure switch lines for condensate. Remove condensate from lines. Check that the condensate lines are located correctly.
4.1.5 Pressure switch does not close due to a low differential pressure across the pressure switch.	**ACTION 1** – Check the differential pressure across the pressure switch. This pressure should exceed the set point listed on the switch. **ACTION 2** – Check for restricted inlet and exhaust vent. Remove all blockage. **ACTION 3** – Check for proper vent sizing and run length. See installation instructions.
4.1.6 Wrong pressure switch installed in the unit, or pressure switch is out of calibration.	**ACTION 1** – Check that the proper pressure switch is installed in the unit. Replace pressure sure switch if necessary.
4.1.7 Miswiring of furnace or improper connections at pressure switch.	**ACTION 1** – Check for correct wiring and loose connections. Correct wiring and/or re-place any loose connections.
4.1.8 Pressure switch failure.	**ACTION 1** – If all the above modes of failure have been checked, the pressure switch may have failed. Replace pressure switch and de-termine if unit will operate.

TROUBLESHOOTING CHART 7.3 (continued) SureLight Control. (*Courtesy of Lennox Industries, Inc.*)

PROBLEM 5: UNIT FAILS TO FIRE IN THE HEATING MODE, COMBUSTION AIR BLOWER ENERGIZES, IGNITER IS ENERGIZED.

Condition	Possible Cause	Corrective Action/Comments
5.1 – Unit operates with a cooling and continuous fan demand. – Combustion air blower energizes with Heating demand. – Ignitor is energized but unit fails to light.	**5.1.1** Check that gas is being supplied to the unit.	**ACTION 1** – Check line pressure at the gas valve. Pressure should not exceed 13" WC for both natural and propane. Line pressure should read a minimum 4.5" WC for natural and 8.0"WC for propane.
	5.1.2 Miswiring of gas valve or loose connections at multi–pin control amp plugs or valve.	**ACTION 1** – Check for correct wiring and loose connections. Correct wiring and/or replace any loose connections.
	5.1.3 Malfunctioning gas valve or ignition control. LED#1–Alternating Slow Flash LED#2–Alternating Slow Flash	**ACTION 1** – Check that 24V is supplied to the gas valve approximately 35 seconds after heat demand is initiated. **ACTION 2** – Replace the valve if 24V is supplied but valve does not open (check for excessive gas line pressure before replacing gas valve). **ACTION 3** – Replace the control board if 24V is not supplied to valve.

PROBLEM 6: BURNERS LIGHT WITH A HEATING DEMAND BUT UNIT SHUTS DOWN PREMATURELY

Condition	Possible Cause	Corrective Action/Comments
6.1 – Burners fire with a heating demand. – Burners light but unit shuts off prior to satisfying T-stat demand. – Diagnostic lights flash the pressure switch code.	**6.1.1** Wrong concentric vent kit used for terminating the unit.	**ACTION 1** – Check vent termination kit installed. 1–1/2" dia. concentric vent (kit60G77) for 50 and 75 inputs and 2" dia. concentric vent (kit 33K97) for 100 &125 inputs.
	6.1.2 Condensate drain line is not draining properly.	**ACTION 1** – Check condensate line for proper vent slope, and any blockage. Condensate should flow freely during operation of furnace. Repair or replace any improperly installed condensate lines.

Code	Condition	Action
LED#1–Off LED#2–Slow Flash **6.1.3** Low pressure differential at the pressure switch.		**ACTION 1** – Check for restricted vent inlet or exhaust. Remove all blockage. **ACTION 2:** Check for proper vent sizing. See installation instructions.
6.2 – Combustion air blower energizes with a heating demand. – Burners light but fail to stay lit. – After 5 tries the control diagnostics flash the watchguard burners failed to ignite code.	**6.2.1** Sensor or sense wire is improperly installed.	**ACTION 1** – Check that sensor is properly located (page 10) and that the sense wire is properly attached to both the sensor and the control.
	6.2.2 Sensor or sense wire is broken.	**ACTION 1** – Check for a broken sensor. **ACTION 2** – Test continuity across the sense wire. If wire or sensor are damaged replace the component.
	6.2.3 Sensor or sensor wire is grounded to the unit.	**ACTION 1** – Check for resistance between the sensor rod and the unit ground. **ACTION 2** – Check for resistance between the sensor wire and the unit ground. **ACTION 3** – Correct any shorts found in circuit.
LED#1–Alternating Slow Flash LED#2–Alternating Slow Flash	**6.2.4** Control does not sense flame.	**ACTION 1** – Check the microamp signal from the burner flame. If the microamp signal is below 0.70 microamps, check the sense rod for proper location or contamination. **ACTION 2** – Replace, clean, or relocate flame sense rod. If rod is to be cleaned, use steel wool or replace sensor. DO NOT CLEAN ROD WITH SAND PAPER. SAND PAPER WILL CONTRIBUTE TO THE CONTAMINATION PROBLEM. **NOTE:** Do not attempt to bend sense rod.

TROUBLESHOOTING CHART 7.3 (*continued*)   SureLight Control. (*Courtesy of Lennox Industries, Inc.*)

PROBLEM 6: BURNERS LIGHT WITH A HEATING DEMAND BUT UNIT SHUTS DOWN PREMATURELY (CONT.)

Condition	Possible Cause	Corrective Action/Comments
6.3 – Combustion air blower energizes with a heating demand. – Burners light. – Roll-out switch trips during the heating demand. – Diagnostic lights flash roll-out failure. LED#1–On LED#2–Slow Flash	**6.3.1** Unit is firing above 100% of the nameplate input.	**ACTION 1** – Check that the manifold pressure matches value listed on nameplate. See installation instructions for proper procedure. **ACTION 2** – Verify that the installed orifice size match the size listed on the nameplate or installation instructions. **ACTION 3** – Check gas valve sensing hose to insure no leaks are present. **ACTION 4** – Check the input rate to verify rate matches value listed on nameplate.
	6.3.2 Gas orifices leak at the manifold connection.	**ACTION 1** – Tighten orifice until leak is sealed. **NOTE:** Be careful not to strip orifice threads. **ACTION 2** – Check for gas leakage at the threaded orifice connection. Use approved method for leak detection (see unit instructions).
	6.3.3 Air leakage at the connections between the primary heat exchanger, secondary heat exchanger, and combustion air blower.	**ACTION 1** – Check for air leakage at all joints in the heat exchanger assembly. Condition will cause high CO2 with high CO. **ACTION 2** – Seal leakage if possible (high temperature silicon is recommended), replace heat exchanger if necessary, tag and return heat exchanger to proper Lennox personnel.

6.4 – Combustion air blower energizes with a heating demand. – Burners light roughly and the unit fails to stay lit. – Diagnostic lights flash watchguard flame failure. LED#1–Alternating Slow Flash LED#2–Alternating Slow Flash	**6.3.4** Insufficient flow through the heat exchanger caused by a sooted or restricted heat exchanger.	**ACTION 1** – Check for sooting deposits or other restrictions in the heat exchanger assembly. Clean assembly as outlined in instruction manual. **ACTION 2** – For G26 gas furnaces, check for proper combustion and flow. CO_2 should measure between 6.0% and 8.0% for NG and between 7.5% and 9.5% for LP. CO should measure below .04% (400PPM) in an air–free sample of flue gases for either NG or LP.
	6.3.5 Burners are not properly located in the burner box.	**ACTION 1** – Check that the burners are firing into the center of the heat exchanger openings. Correct the location of the burners if necessary.
	6.4.1 Recirculation of flue gases. This condition causes rough ignitions and operation. Problem is characterized by nuisance flame failures.	**ACTION 1** – Check for proper flow of exhaust gases away from intake vent. Remove any obstacles in front of the intake and exhaust vent which would cause recirculation. **ACTION 2** – Check for correct intake and exhaust vent installation. See instructions
	6.4.2 Improper burner cross–overs	**ACTION 1** – Remove burner and inspect the cross–overs for burrs, or any restriction or if crossover is warped. Remove restriction or replace burners.

TROUBLESHOOTING CHART 7.3 (*continued*) SureLight Control. (*Courtesy of Lennox Industries, Inc.*)

PROBLEM 6: BURNERS LIGHT WITH A HEATING DEMAND BUT UNIT SHUTS DOWN PREMATURELY (CONT.)

6.5	6.5.1	
– Combustion air blower energizes with a heating demand. – Burners light. – Diagnostic lights flash watch guard flame failure. – NOTE" Unit might go into 60 minute Watchguard mode depending on intermittent nature of sensor signal. LED#1–Alternating Slow Flash LED#2–Alternating Slow Flash	Loose sensor wire connection causes intermittent loss of flame signal.	**ACTION 1** – Check that the sensor is properly located. **ACTION 2** – Check that the sense wire is properly attached to both the sensor and the control. Pay extra attention to the pin connectors.

PROBLEM 7: CONTROL SIGNALS LOW FLAME SENSE DURING HEATING MODE

Condition	Possible Cause	Corrective Action/Comments
7.0	7.1.1	
– Unit operates correctly but the diagnostic lights flash low flame sense code.	Sense rod is improperly located on the burner.	**ACTION 1** – Check the sense rod for proper location on the burner. Properly locate the sense rod or replace if rod cannot be located correctly.

LED#1–Slow Flash LED#2–Fast Flash	7.1.2 Sense rod is contaminated.	**ACTION 1** – Check sense rod for contamination or coated surface. Clean the sense rod with steel wool or replace sensor. DO NOT USE SAND PAPER TO CLEAN ROD. SAND PAPER WILL CONTRIBUTE TO THE CONTAMINATION PROBLEM.

PROBLEM 8: INDOOR BLOWER FAILS TO OPERATE IN COOLING, HEATING, OR CONTINUOUS FAN MODE

Condition	Possible Cause	Corrective Action/Comments
8.0 – Indoor blower fails to operate in continuous fan, cooling, or heating mode.	**8.1.1** Miswiring of furnace or improper connections at control or indoor blower motor.	**ACTION 1**–Correct wiring and/or replace any loose connections. Check for correct wiring and loose connections.
	8.1.2 120V is not being supplied to the indoor air blower or blower motor failure.	**ACTION 1** – Check for 120V at the various calls for indoor blower by energizing "Y", "G", and "W" individually on the low voltage terminal strip. Note that when "W" is energized, the blower is delayed 45 seconds. If there is 120V to each motor tap but the blower does not operate, replace the motor.
	8.1.3 Defective control board	**ACTION 1** – If there is not 120V when "Y", "G", or "W" is energized, replace the control.

TROUBLESHOOTING CHART 7.3 (*continued*) SureLight Control. (*Courtesy of Lennox Industries, Inc.*)

Introduction to Humidifiers

As a heating professional, you have many jobs to perform other than maintenance and repair of heating systems. It is your job to make recommendations to the homeowner on how he or she can use the existing heating system and convert this basic unit into a complete *home comfort package.*

There are three components to this package that you will be called on to recommend to your client. In some cases, you also may have to install these units. These three components are

1. *Humidifiers.* This unit mounts on the heating system and supplies warm, moist air to the home during the heating cycle.

2. *Electronic air cleaners.* This unit mounts in the return air ducting and replaces the air filter. This unit is used to effectively remove dust, dirt, and pollen from the air.

3. *Heat pumps.* A heat pump is installed outside the home and is attached to the existing heating unit. The heat pump supplies heat in the winter and air conditioning in the summer. In some installations, it also can be used to supply "free" hot water.

Chapter 9 will cover the installation and maintenance of humidifiers. Chapter 11 will discuss electronic air cleaners, and I have included an entire section (Sec. 5) on the subject of heat pumps because they are a major part of the heating profession.

There are several types of humidifiers on the market today. Some of these units are used without the aid of the heating system. These "portable" systems do not offer the homeowner the convenience of low maintenance. These units have to be filled with water and sometimes a conditioner. They are noisy and blow cold, moist air into the home. They do not offer the benefit of helping to raise the temperature of the air entering the home.

The other types of humidifiers that will be covered in this chapter are:

1. *Drum type.* This type of humidifier is used in an installation where the heating system is not located near a floor drain. It uses a rotating drum with a filter attached to collect moisture from a reservoir.

2. *Spray type.* In this type of application, a nozzle sprays moisture onto a filter where the warm air blows across this filter to introduce moisture into the warm air stream. This type of unit must be located close to a floor drain as any excess moisture is drained out of the unit.

Both of these units are very effective in introducing moist air into the warm air stream and into the home. They are both controlled by a humidistat that controls the amount of moisture in the home in much the same way that the thermostat controls the temperature in the home.

Figure 8.1 shows the relationship between the static pressure, and the amount of moisture that will be introduced into the home based on the supply air temperature of the heating system. In Chap. 6 you learned how to measure this temperature. As you will see from Fig. 8.1, the *higher* the static pressure (water pressure), the *higher* the supply temperature, the more moisture that will be introduced into the home. It is very possible to introduce 10 to 20 gallons of moisture into the home.

You may be asked by the homeowner, "why is it important to have a humidifier"? The answer is simple. Ask the homeowner these questions:

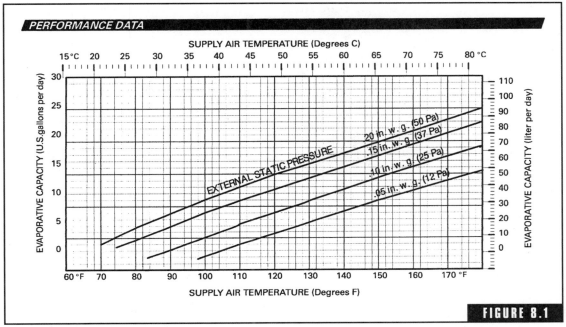

Static pressure chart. (*Courtesy of Lennox Industries, Inc.*)

1. Do you ever wake up in the morning, and your nose is very dry?

2. Do you notice that there is a lot of static electricity in the home in the winter?

3. Does your family have more than their share of colds in the winter?

4. Do you have any furniture that you have noticed the glue joints drying out?

If the answer to any of these questions is "yes," they need to have a humidifier installed on their heating system. The amount of moisture that will be introduced into the home will make it a much more comfortable place to live. Moist warm air will actually raise the temperature of the air entering the home. The best example to use for the homeowner is to ask them if they notice that on a warm summer day, when the humidity is high, it seems to be much warmer than on a day where the temperature is the same, but there is no humidity in the air? This should help to make your point to the homeowner that this is the same principal that applies here.

By having a humidifier installed on their heating system you are helping the homeowners to protect their investment in their families health. They will notice a drop in the amount of dry heat health problems that their family has during the winter months. You will also be helping them to protect the investment they have in their home by eliminating the problems such as static electric build up, that occur when dry static heat is introduced into the home. The final result will be that the homeowner will have made a wise investment and chosen to make their home more comfortable during the winter months as well.

Installation and Maintenance of Humidifiers

You learned about the types of humidifiers and the benefits to homeowners of having a humidifying unit installed on their heating system in the preceding chapter. Now you will learn how to install and maintain these units.

Figures 9.1 through 9.3 show the three types of humidifiers that will be covered in this chapter. These would be the drum type, the spray type, and the steam type. All these units have the same result, but they operate slightly differently. Installation is slightly different depending on which unit you are installing.

Figure 9.4 shows the basic unit layout of the drum- and spray-type humidifiers. You will notice that the endplates can be removed to change the direction of the airflow depending on which side of the unit the return airflow is coming from.

To begin, look at Figs. 9.5 to 9.8 to determine which type of heating system you will be installing the unit on. Notice that the humidifier

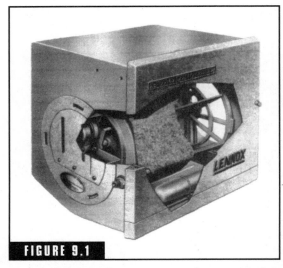

Drum-type humidifier. (*Courtesy of Lennox Industries, Inc.*)

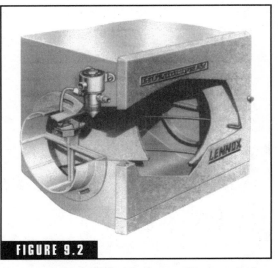

Spray-type humidifier. (*Courtesy of Lennox Industries, Inc.*)

itself mounts on the *return air side* of the heating unit, and the collar is mounted on the *supply air side* of the unit. This is done so that the warm air is blown over the filter medium so that the moisture can be introduced into the warm airstream.

Before you begin installation of the unit, you should have made some basic observations:

1. Is there a drain located close to the heating system? This will determine which unit you will install.

2. Is there a water line located close to the heating system? You will want to make the water connection as close to the unit as possible. You also should notice if the water lines are plastic, copper, or galvanized. This will make a difference when you install the saddle valve later in the installation.

3. The location on the heating system to mount the unit. You will want to make sure that you have enough room to install the unit.

4. The location of the transformer. Will you mount it inside the wire cabinet or outside?

5. The location of the humidistat. Will you mount it on the wall or on the duct?

After you have answered these questions, you are ready to begin the installation.

Installation

First, turn the power off to the unit; then remove all the parts from the box, and check them with the parts list. Make sure that you have all the parts before you begin. If you are missing any parts, get the parts needed before you begin the installation.

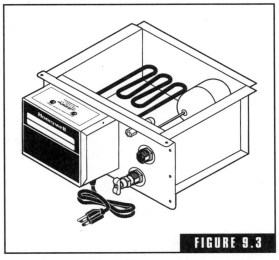

Steam-type humidifier. (*Courtesy of Honeywell, Inc.*)

Once you are convinced that everything is in place, use the template to make the location on the return air duct where you will mount the box. It is important to mention here that you need to make sure that the unit will mount as level as possible. This not only makes the installation look more professional but also will help with the drainage of excess water.

Once you have the template in place, drill the mounting holes in the locations indicated on the template. Next, cut the hole for the center of the unit. This will be the area in front of which the filter medium will be located to allow the moisture to enter the warm airstream. Start two screws in the upper mounting holes, and leave them extended. Mount the unit on these two screws, and check for proper alignment. The unit should be centered on the hole that you cut in the duct work. Make adjustments as needed to center the unit. Tighten the two upper screws, and install the bottom screws to hold the unit in place.

Remove the end cap for the end of the unit that will face the supply air side. Since this is a universal unit, you may have to reverse the end caps so that the spray nozzle (on a spray unit) or the motor assembly (on a drum unit) is located on the opposite end of the unit.

Locate the area on the return air duct where you will be mounting the damper starting collar. This should be located in a straight line with the unit, since you will be using the flexible duct to connect this

SPECIFICATIONS

Model No.	WS1-18
Nozzle size – U.S. gph (L/hr)	3.00 (11)
Type of nozzle	Delavan WD8
Nozzle spray angle	90°
Filter media cut size – in. (mm)	13-7/8 x 12-1/2 x 1/2 (352 x 318 x 13)
Water supply connection – in. (mm)	1/4 (6) compression
Drain connection – in. (mm)	3/4 (19) I.D.
Unit operating watts	10
Electrical characteristics	24v — 60hz— 1ph
Shipping weight – lbs (kg) (1 pkg.)	11 (5)

* Rated at 18 U.S. gallons (68 L) per day evaporation; 140° F (60°C) supply air temperature .20 in .w.g. (50 Pa) static pressure and 30% return air humidity.

DIMENSIONS — inches (mm)

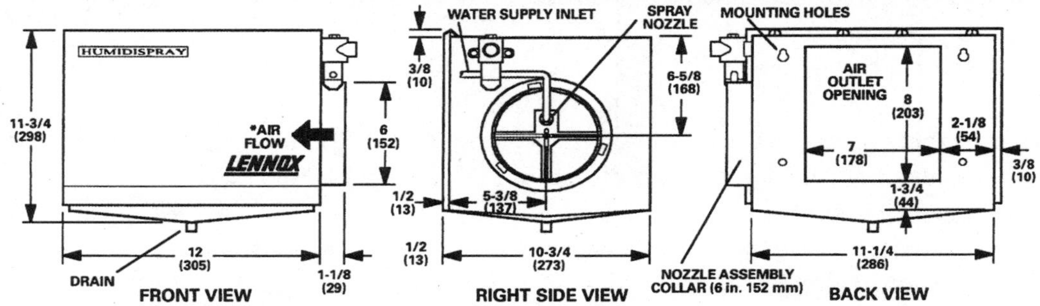

*NOTE — Right hand inlet air shown. Relocate inlet collar and motor assembly to obtain left hand inlet air.

FIGURE 9.4

Humidifier layout. (*Courtesy of Lennox Industries, Inc.*)

collar with the unit. You do not want to make any more bends in this line than are absolutely necessary, since every bend reduces the amount of airflow.

Hold the collar on the return air duct, and trace the inside circle of the collar on the duct. Cut this opening out, and fit the collar in place. Make any additional cuts needed to this circle to make the collar fit snugly. After you have a proper fit, reach inside the collar and bend the tabs down against the duct to lock the collar into place. Slide one large clamp over one end of the flexible duct, and slide this end over the collar. Slide the clamp into place so that the clamp is over the flexible duct and collar. Tighten this clamp in place, taking care not to overtighten the clamp. Do not attach the other end of the duct to the unit at this time.

Locate the water line that you will be using to supply water to the unit. If the connection is copper or plastic, this should be no problem. If the connection is galvanized, you may have a problem with the saddle valve installation, since this is a very hard type of pipe. It is important to remember that once you begin to install the saddle valve, you are committed to finish. Once you pierce the line, you must finish. You also may have a lower water flow from this type of pipe because it is used in older homes and there could be a buildup of corrosion in the line.

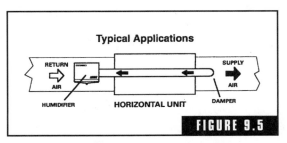

Horizontal installation. (*Courtesy of Lennox Industries, Inc.*)

Remove the saddle valve parts from the bag. You should check the assembly instructions on the bag to make sure that you have all the parts needed for the installation.

Place one of the bolts through one of the holes in the saddle valve, and attach a nut to the other end. Do not attach the other bolt at this time. Place the rubber gasket over the end of the valve. Locate the place on the water line that you have selected for the installation. Put the saddle valve in place over the water line, making sure that the outlet on the valve is facing the humidifier. Attach the other bolt and nut on the saddle valve. Center the valve over the water line, and tighten the bolts until the unit is secure to the line.

Measure the distance between the saddle valve and the connection on the humidifier. Cut a piece of the $1/4$-in plastic water line slightly longer than your measurement. This is the line that will be attached to the saddle valve and the humidifier. Slide one of the retaining nuts over the end of the

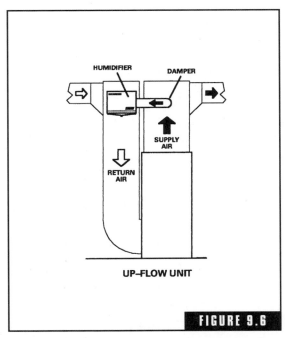

Upflow installation. (*Courtesy of Lennox Industries, Inc.*)

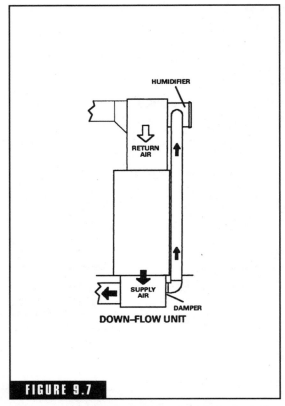

HUMIDIFIER

RETURN AIR

SUPPLY AIR

DAMPER

DOWN–FLOW UNIT

FIGURE 9.7

Downflow installation. (*Courtesy of Lennox Industries, Inc.*)

plastic tube. Now slide one of the compression fittings over the end of the plastic line. Insert one of the brass inserts into the end of the plastic line, if so equipped. Insert the end of the plastic tube into the saddle valve, and push it in as far as it will go. Slide the compression fitting and nut up to the valve, press the compression fitting into the saddle valve, and tighten the nut in place. This will cause the compression fitting to seal itself to the plastic line. Repeat this same process on the other end of the plastic line, and attach it to either the nozzle assembly on a spray-type unit or the solenoid on a drum-type unit. Do not turn the water on at this time.

Remove the transformer from the box. Mount the transformer in the location that you chose. You must follow the instructions included with the unit when wiring the transformer into the blower circuit. On newer units, there will be an accessory terminal that can be used to install the transformer. If no such terminal is available, you will need a current-sensing relay (on Lennox units this is Part Number 26G12) so that it will control the solenoid water valve on the humidifier. On older units, the transformer must be wired with the fan control circuit but not attached to the wiring that directly supplies power to the blower motor.

Next, install the humidistat in the location that you have chosen. Wire the humidistat with the transformer so that the humidistat is the switch that controls the operation of the solenoid valve to allow water to either enter the reservoir (in a drum-type unit) or supply water to the nozzle (in a spray-type unit). Follow the wiring instructions that come with the unit.

Secure the blank end cap on the unit, and install the foam filter medium onto either the drum or frame. Install these filters in the unit, and secure them into place.

With both ends of the plastic water line in place and secured, turn the handle on the saddle valve clockwise to pierce the water line. Be sure to screw this handle all the way in until it will not go any farther. This not only makes a hole in the water line but also is the shutoff for the valve. There should be no water flowing at this time. Now, turn the handle counterclockwise to allow the water to flow to the unit. Turn this handle slowly, and watch for water leaks. Tighten any bolts and/or fittings necessary to stop any leaks. Once you have all the leaks stopped, slide the remaining clamp over the end of the flexible duct, and secure in place over the open end of the humidifier. Secure this flexible duct into place.

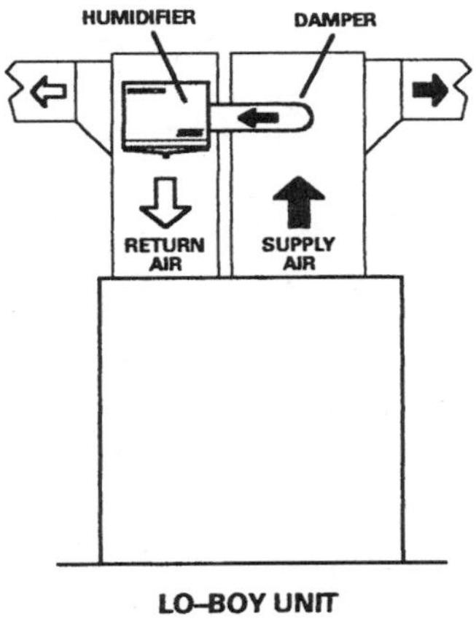

If you have installed a drum-type unit, you should have water flowing into the

FIGURE 9.8

Low-boy installation. (*Courtesy of Lennox Industries, Inc.*)

reservoir. Make sure that the float valve is working properly. When the water reaches the mark on the inside of the unit, the float should close and stop the flow of water to the unit. If not, adjust the float so that the water stops at the proper point. There should be enough water in the reservoir for the bottom of the filter medium to be able to pick up this moisture when it rotates during normal operation. Now attach the drain hose to the bottom of the unit, and secure it with the clamp provided. On spray-type units, run the drain line to the floor drain, cut to length as needed. On drum-type units, run the drain hose to a floor drain if available. It is a good idea to inform the homeowner at this point where the saddle valve is located so that it can be shut off if the float sticks and there is a need to shut the water off to the unit.

Once all the installation is complete, it is time to check the operation of the unit. The unit should operate only when the blower is running. Since some homes are

QUICK»TIP

If no drain is available, use a bucket or some device that can be placed under the unit to catch the excess water until the valve can be shut off in the event that the water does not shut off automatically.

equipped with central air-conditioning units or heat pumps, it is important to show the homeowner how to shut off the humidifier in the summer. Since these other home comfort units require the use of the heating system blower to operate, if the humidifier is not turned off in the summer, it also will run while these units are operating. Instruct the homeowner to turn the humidistat all the way down (counter-clockwise) to shut the unit off. Do not allow the homeowner to simply turn the water supply off, since this may cause damage to the solenoid valve because there will be no water supply to cool this device.

Turn on the power to the unit, and turn the thermostat up if the unit did not start. Watch the operation of the humidifier. If the unit begins to operate as soon as the heating system starts, you need to check the wiring of the transformer and humidistat, because they are not wired properly. Turn the power off to the unit, and make the corrections. Turn the power back on, and check the operation again. When the blower begins to blow, the unit should begin to operate. With a drum-type unit, make sure that the drum is rotating freely and that there is enough water in the reservoir for the filter medium to pick up this moisture. On a spray-type unit, make sure that the nozzle is spraying properly onto the filter medium. When you are sure everything is operating properly, replace the front cover on the unit, and tighten the screws that hold this front cover in place. Set the humidistat to about 20 percent as a good starting point. Show the homeowner how to operate this control so that he or she can adjust the amount of moisture in the home once the unit has operated for several days. It will take that long to replace the moisture in the home that has been lost.

A steam-type humidifier is much easier to install than the other two types just discussed. Since this type of humidifier is more of a self-contained unit, much of the installation involved with the other types of units is not necessary. Figure 9.3 shows a steam-type humidifier.

To begin, look at Figs. 9.5 to 9.8 to determine which type of heating unit you are working on. This type of humidifier is one piece, so there is no need to attach the flexible duct to this unit.

Locate a good mounting spot on the *hot air duct*. Use the template to get the correct mounting arrangement. You will need to cut a hole in the hot air duct to allow the unit to be installed. Use a sharp tool to make a hole in the center of the template. Use your tin snips to begin cutting out the opening. You should begin from the center and cut out to the four corners first. Once this is done, make the horizontal and vertical cuts. When you are finished, make sure that there are no sharp edges on the cuts. If there are, use a file to remove any burrs that were left. You also can use tin benders to bend the edge over to make a neat-looking, smooth edge.

Once you have made the opening, test fit the unit in place. Since this type of unit mounts inside the duct versus mounting to the duct, the opening size is more critical. Make any further cuts to allow for smooth installation of the unit into the duct. Be careful not to make the additional cuts too large so that there are gaps between the edge of the unit and the duct. Since you will be sealing this unit to the duct, you do not want gaps that could cause air leaks.

Once the opening is the proper size, you will need to drill the mounting holes. Use a drill bit of the size specified in the instructions. You can use the markings on the template to make the holes, but I prefer to hold the unit in place and then drill the holes. The reason for this is that if you had to make additional cuts during the last part of the installation, you may have moved the location of these holes. Hold the unit in place, and drill the first mounting hole. I suggest that you start with one of the upper mounting holes. Once you have the first hole drilled, use one of the mounting screws to secure the unit in place. Do not overtighten the screw. Place a small level on the top front edge to make sure that the unit is level. This is one of the critical steps in this type of installation. Since this unit holds water in a tank, if the unit is not level, the float may cause the water to overflow into the heating unit. This could cause damage to the heating unit, so making sure that the unit is installed level side to side is critical.

Once you have the unit level, drill the remaining mounting holes. Once this is done, remove the mounting screw, and remove the unit from the duct. Apply the foam sealing tape to the location specified on the template. This will help to seal the unit to the duct to avoid air leaks. This is why it is so critical to avoid making the mounting hole too large, since the sealing tape may not be able to stop all the air leakage from around the unit.

Now that the sealing tape is in place, slide the unit in the mounting hole, and secure in place with the mounting screws. Once all the screws are in place, check the unit again to make sure that it is level.

Make any adjustments necessary. When the unit is level and all the screws are in place, you are ready to install the water line.

Most units will include a saddle-type valve to mount to the water line, some plastic tubing, and fittings. If your unit is different, follow the instructions that come with the unit for installation of the water line.

On units that have these components, first assemble the saddle valve. Use the long bolts, and place one in each of the bracket holes on the valve body (the bracket with the T handle on it). Now locate the water line on which you will install the saddle valve. This should be located as close to the unit as possible. Copper is the best type of line on which to install the saddle valve, but it will work on any type of line. First, place the rubber gasket in place on the valve body. This gasket mounts between the valve body and the water line. It is the seal that is used to prevent leaks. Now take the lower bracket and slide the bolts through the mounting holes. Slide the bracket up to meet the lower part of the water line. Thread the nuts on the bolts, and make them finger tight to hold the saddle valve in place. Make any final adjustments to the valve to make sure that the rubber gasket is in place and the valve is seated properly on the line. Both the valve assembly and the lower bracket have U-shaped indents that mount around the pipe. These are used to locate the valve properly on the pipe.

Once you have made the final adjustments to the valve, tighten the valve to the water line. Do not overtighten the valve to the line. You do not want to crush the line by making the valve too tight on the line, or leaks can occur.

Take the plastic tubing that came with the unit. There also should be two brass nuts, two brass inserts, and two compression fittings packed with the valve. Begin by sliding one nut onto the water line. Slide a compression fitting onto the line. Place one of the brass inserts into the end of the tubing. This will keep the line from collapsing. Slide the end of the tubing into the opening in the valve. Slide the nut and compression fitting up to the threads on the valve. The compression fitting should seat itself in the opening. Tighten the nut on the threads. This will cause the compression fitting to compress onto the water line, creating a seal between the water line and the valve.

Measure the distance from the valve to the threaded fitting on the humidifier, and cut the line to length. Make sure that you leave enough slack so that you can make the installation as neat as possible. You want to attach the water line with cable staples to the floor joist or beam so that the line does not hang. Once this is done, install the water line to the humidifier in the same fashion that you did to the valve with the nut, compression fitting, and insert. Tighten the water line to the threaded fitting on the float assembly of the humidifier.

You are now ready to activate the saddle valve. This involves turning the T handle all the way in to pierce the water line. This then acts to close the valve and stop the flow of water. By turning the handle in the opposite direction, you open the valve. One point to remember is that once you close the valve for the first time, you are making a hole in the water line that you cannot repair, so it is critical that you make sure that the saddle valve is installed properly prior to closing the valve for the first time.

Once you have closed the valve as far as you can turn the handle, open the valve slowly. Check for leaks as you open the valve. If there are leaks, tighten the fittings at the point of the leak until the leak stops. Where I have found most leaks to occur is at the nut located on the T handle. Make sure that you have the proper size wrench available when you open the valve so that you can tighten this nut.

Once all the leaks (if there were any) are stopped, listen for the water to be entering the humidifier tank. There is a float valve in the tank that should shut off the water when it reaches the proper level. These units are equipped with a low water level cutoff switch and overflow protection. This will help to keep the unit from malfunctioning. Since these units do not have to be wired to the transformer and blower circuit, all that needs to be done for power is to connect the unit to a standard 110-V ac outlet. However, before plugging in the unit, you need to install the humidistat. This is the unit that controls the amount of moisture in the home. This unit is mounted on the return side of the heating system. Follow the wiring instructions that came with the unit so that the humidistat is installed properly for the unit you are working with.

Plug the unit into an outlet, and turn the humidistat up until the unit begins to operate. As stated earlier, this type of unit does not depend on the forced air heating system to cycle before it will operate. This type of unit has a built-in heating unit that will heat the water into a vapor so that it can work independently of the heating demands. The warm, moist air will gravity feed into the home, or the blower control can be set to "manual" so that the blower operates all the time. This will help to better distribute the moisture when the heating system is not calling for heat.

Once you have checked the operation and you are satisfied that the unit is operating properly, the installation is complete. Before leaving, though, make one last check to make sure that there are no leaks and that there is water running out of the drain hose on spray-type humidifiers. Wait until the heating unit has run for a complete cycle to make sure that the humidifier will shut off when the blower shuts off. With a steam-type unit, turn the humidistat down below the humidity level to make sure that it will shut off properly. If there are no leaks and the operation is satisfactory, the installation is complete, and you can now move on to the next section on maintaining these units.

Maintenance

Humidifiers, like any part of a total comfort system, require maintenance from time to time. The amount of maintenance that is required is small, but it is required to keep the humidifier in top operating condition.

Humidispray unit

Turn off the power to the unit. Loosen the screws that hold the front cover of the unit. When you remove the front cover of the unit, observe the condition of the unit. Do you see a large buildup of lime and minerals on the filter medium and cabinet? If you do, this is an indication that the homeowner has a hard water situation. This may require that the unit be serviced more than once a year to keep the proper amount of humidity in the home. You should talk to the homeowner and recommend that he or she invest in a *water conditioning unit.*

A water conditioning unit will make the water "soft," which will help to reduce the amount of minerals in the water. This will help more than just the maintenance required on the humidifier and will save the homeowner money in laundry soap and other household products.

Remove the filter frame from the unit. Remove the filter medium from the frame, and remove any buildup on the frame. This can be done very quickly with a solution of vinegar and water. This will remove the lime and deposits from the whole unit. If you do not have this with you, you will need to scrape the deposits from the frame, and wash the frame off with plain water.

Remove the flexible duct from the unit by loosening the clamp, and slide it off the unit to expose the nozzle assembly. Remove the nozzle assembly from the humidifier. Hold the assembly with a wrench, and use another wrench to remove the nozzle. Be sure to replace the nozzle with one of the same size.

Use the vinegar and water solution to clean the inside of the cabinet to remove the lime and mineral buildup that will be there. Clean the unit completely so that it will operate properly.

Replace the filter medium with a new filter, and install the frame into the unit. Reinstall the nozzle assembly into the unit, and replace the flexible duct and tighten the clamp into place.

Pour some water down the drain to make sure that it is clear. Watch the end of the hose that is located at the floor drain, and make sure that the water comes out the end. If you are not getting a good flow or no flow, you may have an obstruction in the line that must be cleared or the hose replaced. Depending on the length of the drain line, you may be able to run a wire down the line to clear it. If the line is too long, try to power flush the line by removing it from the unit and using a garden hose with a spray nozzle attached to flush the line clear. I recommend that you take the line outside to do this, since there will be dirty water coming out that you do not want in the client's home. It is much better to take the time to do this outside than to have an accident happen in the home.

If the line cannot be cleared, replace it with a new one. The cost is inexpensive, and this will keep the unit draining properly in the winter.

Perform a run check on the unit once you have installed the drain hose. Turn the power back on to the unit. Turn the thermostat up to call for heat, and wait for the blower to start. When the blower comes on, the humidifier should begin to operate. If not, change the setting on the humidistat until the unit starts. Make sure that you have a good spray coming from the nozzle and that it is spraying properly onto the filter. Replace the front cover, and allow the unit to run until you see water coming from the drain hose. Once you have established a good drain, turn the thermostat back down, and allow the unit to complete the cycle.

Humididrum Unit

The maintenance on a humididrum unit is very similar to that on a spray unit except that you do not have a nozzle and there should be no water coming from the drain. If you do, you will need to examine the float to see if it is faulty or sticking as a result of a buildup of minerals or lime.

Turn the power off to the unit. Remove the front cover from the unit. Observe the condition of the unit for mineral deposits. If there is a large buildup of these deposits, again recommend to the homeowner that he or she have a water conditioning unit installed to soften the water. You may need to perform maintenance on the unit more than once a year if the condition of the water is such that large deposits of lime and minerals form on the filter medium and cabinet.

Remove the drum by lifting up on the end opposite of the motor unit and then sliding the drum off the motor unit. Remove the filter medium from the drum, and clean the drum with the solution described in the preceding section. Again, if you do not have this solution with you, you will need to remove all the deposits from the drum and unit by scraping them off.

Replace the filter medium with a new filter. Clean the inside of the cabinet so there are no signs of deposits. Check the operation of the float by moving it gently up and down. If you see lime buildup on the float and arm assembly, this will need to be cleaned. It is very important with this type of unit to make sure that the float is operating properly and free from anything that will keep it from maintaining the proper water level in the unit.

Reinstall the drum in the unit. Turn the power back on, and set the thermostat to call for heat. Once the blower comes on, the drum should begin to turn. If it does not run, turn the humidistat up until the unit comes on. Turn the thermostat back down, and allow the unit to complete the cycle. Replace the front cover on the unit.

Steam Unit

The maintenance of this type of unit is very low because it has a timed flushing cycle that reduces the amount maintenance needed in a hard water installation. Before performing any maintenance on this unit, unplug the line cord from the outlet. All that needs to be done is to remove the unit from the heating system and clean out the tank to remove any deposits. You also will want to remove any deposits from the heating element, since lime deposits are a good insulator and will keep the unit from operating properly. Check the valve on the float for signs of lime buildup as well. Any buildup on the float valve can cause the valve to not close all the way, allowing more water to enter the tank than is needed.

Is an Electronic Air Cleaner Right for You?

Another component of the total comfort package is an *electronic air cleaner*. Units such as this are designed to electronically remove particles from the air that would not be removed by normal filters installed in your heating system.

As a heating professional, it is part of your job to make recommendations to the client based on the information that he or she supplies or the observations that you make. You also need to be prepared to answer the questions that your client may ask you about these units. I have mentioned this before in previous chapters because I feel that it is important that you be aware of the needs of the client. Clients may not be aware of all the comfort items that are available to them to make their quality of life better.

I know it sounds kind of strange, that a heating system can improve a person's quality of life, but it truly can. In the preceding chapter on humidifier installation and maintenance, you learned that if you listened to your client's needs in terms of the problem he or she was having with dry noise and static electricity in the home, you solved this problem by installing a humidifier. You also may have observed that the water condition in the home needed attention. If the homeowner took your advice and has a water conditioner installed, his or her quality of life has improved. This is what I am talking about here. You have made the home a more comfortable place to live.

In the case of an electronic air cleaner, the improvement in quality of life will be that the amount of dust and pollen in the air of the home will be reduced. Some people have a problem in the summer when the dogwood trees are in bloom. Some people cannot even leave their homes at these or other times of the year because their allergies are so bad that they are miserable. This is where the installation of an electronic air cleaner will help.

Electronic air cleaners work when the blower is turned on either during the heating/cooling cycle or when the blower is set for continuous operation. It is possible to have the blower running in this mode with the air cleaner installed and to have the air in the home cleaned continuously. This will be a great benefit to clients who have a problem with asthma and allergies in the summer.

A normal heating system filter has an efficiency rating of only about 10 percent. This means that 90 percent of the pollen, mold spores, etc. continue to enter the home. An electronic air cleaner, on the other hand, will remove up to 90 percent of these same particles. This is done without a large pressure drop in the system. Imagine what this will mean to your client who has these problems—that 90 percent of the particles that cause him or her to be miserable in the summer will be removed!

Figure 10.1 shows the efficiency of electronic air cleaners in removing airborne particles from the home. As you can see, some of these particles are very small, some being as small as 0.01 μm.

Electronic air cleaners can remove dust, dirt, and microscopic particles because of the high amount of filtration they employ. Each unit uses duel prefilters that are installed in the unit ahead of the electronic filters. These prefilters trap as much of the dust and dirt as possible. Any particles that pass through these prefilters are then directed through a solid ionizing screen that places a positive charge on them. These positively charged particles then pass into the collection plates that have a

AIR CLEANING EFFICIENCY

On the average, an electronic air cleaner will remove fifteen (15) times as much dust, dirt, lint and mold spores from the air as an ordinary furnace filter. And, on smaller particles, the percentage removed vs. standard filters is significantly greater.

An electronic air cleaner will remove airborne particles as small as 0.01 microns in diameter. The chart below lists sizes of common airborne particles trapped and removed from recirculated air by electronic air cleaners.

Types of Airborne Particles	Particle Size (*Microns)
Pollen	10.0 to 100.0
Tobacco Smoke	0.01 to 1.0
Cooking Smoke	0.02 to 1.0
Household Dust	0.01 to 300.0
Mold Spores	10.0 to 30.0
Atmospheric Dust	0.01 to 1.0
Insecticide Dust	0.40 to 10.0
Coal Dust (Soot)	1.0 to 100.0

*One micron = 1/25,400th of an inch.
Particles 10 microns and larger are visible to the naked eye.
Particles 10 to 0.1 microns are visible with microscope.
Particles below 0.1 microns are visible with electron microscope.

FIGURE 10.1

Air cleaner efficiency. (*Courtesy of Lennox Industries, Inc.*)

negative charge on them. This negative charge attracts the particles and holds them until the cells can be removed and cleaned. If the client needs even more filtration because there are smokers in the home, cooking odors, etc., charcoal filters can be installed on the opposite side of the unit from the prefilters to maximize the amount of filtration.

Figure 10.2 shows the efficiency rating of these units based on the size of the heating system in which they are installed. As you can see, with an air volume of 1600 cfm (cubic feet per minute), you can achieve an efficiency of 95 percent with a Model EAC12-14 from Lennox Industries.

As you have seen in this chapter, anyone who has allergies in the summer should have one of these units installed in his or her home so

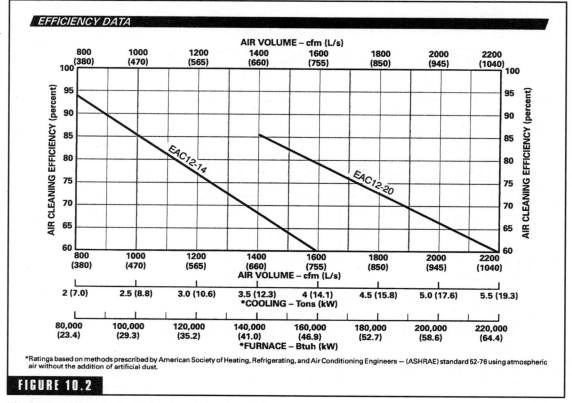

EFFICIENCY DATA

*Ratings based on methods prescribed by American Society of Heating, Refrigerating, and Air Conditioning Engineers — (ASHRAE) standard 52-76 using atmospheric air without the addition of artificial dust.

FIGURE 10.2

Air cleaner data. (*Courtesy of Lennox Industries, Inc.*)

that his or her quality of life can be improved. An electronic air cleaner is also a great addition to the heating system for clients who have smokers in their home and want to keep the home smelling fresh. It is also great for a client who lives on a nonpaved road, where dust is a factor in the dry summer months. By installing an electronic air cleaner and keeping the windows closed, the client can trap up to 90 percent of the dirt particles that normally would be in the home. This would be a large time saver in not having to dust the home as often.

In Chap. 24 we will talk about the final part of the total comfort package—heat pumps. With the addition of a heat pump, it is possible to keep the home warm in the winter, cool in the summer, and virtually dust- and pollen-free. This is a great help to clients who need this type of protection from the elements.

Installation and Maintenance of Electronic Air Cleaners

Installation

Installation of an electronic air cleaner should be done at the time a new heating system is installed. This is the simplest installation. However, it is possible to install an electronic air cleaner at any time. This will involve modifying the duct work to make room for the air cleaner. I do not recommend that a homeowner attempt this procedure himself or herself. This should be left to the heating professional who has access to a sheet metal shop where any additional components can be made if needed.

The first thing that needs to be done is to decide on the location of the unit. Figures 11.1 to 11.5 show the typical locations for these units depending on the type of heating system in the home. It is important to understand that the location of the unit must be in the return air ducting, where the temperature of the air entering the

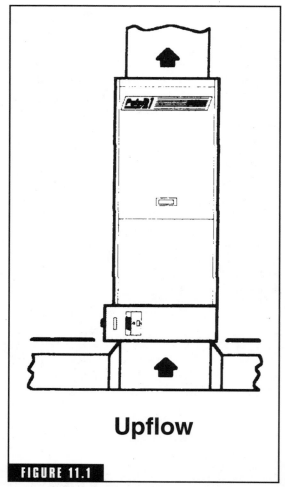

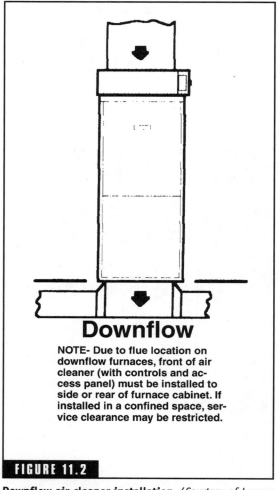

Upflow air cleaner installation. (*Courtesy of Lennox Industries, Inc.*)

Downflow air cleaner installation. (*Courtesy of Lennox Industries, Inc.*)

electronic air cleaner will be between 40 and 85°F. If you are not sure if this is the case, you should place a thermometer in the return air duct above the location where you would like to do the installation. Set the thermostat to call for heat, and watch the temperature reading on the thermometer after it has reached the most stable reading. If the temperature is in the recommended range, you may begin the installation. If the reading is outside the recommended range, find another location in the return air duct that is within the correct range of operation.

You should not install an air cleaner in an area where 100 percent outside air will be running through the unit. You also should not locate the air cleaner in a location that is upstream from the humidifier, since this will decrease the efficiency of the air cleaner. Once you have the unit installed, the best operation is achieved with constant blower operation.

Measure the size of the unit that you will be installing to get the proper dimensions for the opening in the duct work. Figure 11.6 shows the dimensions of one such unit.

Most units are made for universal installation. By this I mean that you do not have to worry about the direction of airflow because the prefilters and electronic cells can be reversed after installation. It is important to note at this point that there are arrows on the prefilters and electronic cells that show the airflow direction. You must make sure that you install these components so that the air is flowing in the proper direction, or the unit will not operate properly.

Once you have decided on the location and have the proper size opening cut, slide the unit into place. Secure the duct work in place around the unit so that there will be no air loss. There can be no openings that will allow any air to enter the unit for any other source than the closed duct work, or the efficiency of the unit will be greatly reduced.

Turn the power off to the unit. You will need to wire the unit in series with the blower circuit. Some heating units have auxiliary strips in the panel that can be used for these types of units. If you have this option available, wire the unit into this strip. Make sure that you have checked local building and wiring codes to make sure that you are in compliance with them. Most states have a code the

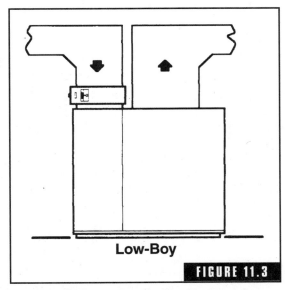

Low-Boy

FIGURE 11.3

Low-boy air cleaner installation. (*Courtesy of Lennox Industries, Inc.*)

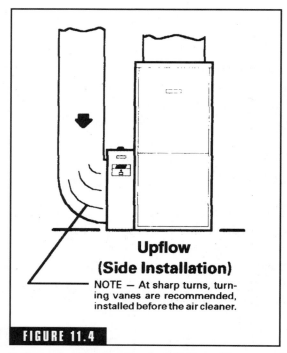

Upflow
(Side Installation)

NOTE — At sharp turns, turning vanes are recommended,
installed before the air cleaner.

FIGURE 11.4

Upflow air cleaner side installation. (*Courtesy of Lennox Industries, Inc.*)

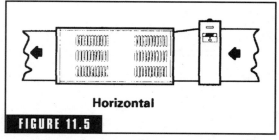

Horizontal

FIGURE 11.5

Horizontal air cleaner installation. (*Courtesy of Lennox Industries, Inc.*)

requires that you place all wiring in conduit. This is always a good idea so that the
wires are protected at all times.

If you do not have an auxiliary strip
available, you will need to wire the unit
in series with the blower circuit so that it
will come on and operate only when the
blower is running. Do not wire this unit
directly to the hot wire of the blower
motor. There are some cases where the
motor can cause the voltage to increase
when it is running. You will need to wire
the unit with the control circuit that
allows power to the blower when either
the fan control calls for the blower or the
manual setting is selected, allowing the
blower to operate separate from the fan
control. Check the wiring diagram on the
unit for this circuit.

Once you have the unit wired to the
proper circuit, you will need to attach
the wires to the electronic air cleaner.
Most units have several "knockouts"
located on the junction box on the unit.
Find the best location for the wiring to be
routed to the unit, and remove the knock-
out plug that is closest to this location.
Install the necessary connector in this
hole to attach to the conduit, and secure
the conduit in place.

Check the diagram for the model of electronic air cleaner that you are installing,
and attach the wires from the blower circuit to the proper wires in the
junction box.

Slide the prefilters into place, noting the airflow. Slide the electronic cells into place, also noting the airflow. If your client has
ordered charcoal filters as well, install these filters on the opposite end

of the unit from the prefilters. Attach the front door in place. You are now ready to check the operation of the unit.

Turn the power to the unit back on, and set the blower control to the manual setting so that the blower will run in a continuous mode. Turn the power switch on the electronic air cleaner on to start the cleaning operation. You may hear arching sounds coming from the unit. This is normal, since any particles that are large enough will cause the charged dc voltage wires to arch, breaking down the size of the particles so they can be trapped on the plates. This is normal operation for an air cleaner.

If the air cleaner does not come on when the blower is running, you will need to check the wiring to make sure that you have wired the unit to the proper circuit. The unit also should shut off when the

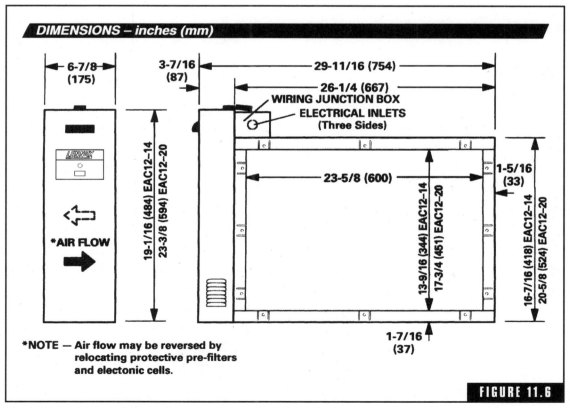

Dimensions of the air cleaner. (*Courtesy of Lennox Industries, Inc.*)

blower is not running. Make the necessary adjustments to the circuit until the proper sequence of operation is achieved.

Maintenance

Maintenance of electronic air cleaners is fairly simple. It involves removing the prefilters and electronic cells and cleaning them. Caution must be exercised because there is a static electric charge that builds up on the grids of the electronic cells that must be discharged before the cells can be removed or the front cover removed on some units. On these units, there will be a button located on the front cover that you press to discharge the cells prior to removal. Make sure that the power to the unit and the heating system is turned off prior to discharging the electronic units. Press the button several times to discharge the units.

Remove the front cover to the unit, and remove the prefilters and electronic cells. Observe the amount of dirt that is on the prefilters and electronic cells. There should be a fair amount of buildup on these items if the unit has not been serviced in some time. If the cells and prefilters are fairly clean, you should ask the homeowner when the last time was that the unit was serviced. If it has been over 6 months, you need to check the operation of the unit prior to cleaning to make sure that it is working properly. Make sure that the unit comes on when the blower is running. If not, check the wiring to make sure that all the connections are correct and tight.

If the connections are fine, turn the power off and discharge the unit. If there is no charge builtup, remove the front cover and check the cells to make sure that they are installed properly. Most often the cells are not installed properly, and they are either facing the wrong way or not making a connection.

There are several ways to clean these units, but I have found the best way is to take them to a manual car wash where you have access to soap and water spray. You can clean these units using a garden hose, but you will not get the same results. Once these filters are cleaned, they must be allowed to dry completely before they can be used. This can be done simply by placing them back in the unit and instructing the homeowner not to run the unit for at least 1 hour to allow the cells to dry. Once the cells are dry, the unit can be returned to normal operation. Make sure to turn the power to the heating system and air cleaner back on before you leave the client's home, or you may be making a return trip to do this later in the day.

Other Options

If the homeowner does not want to go to the full expense of installing an electronic air cleaner at this time, there are some options to explore. One such option is called a *media air cleaner*. It is installed in the same as an electronic air cleaner, except there are no wires to attach. Instead of using an electronic filter, this air cleaner uses a high-efficiency filter. The filter slides into the unit like an electronic cell would, but it is made of a replaceable filter medium.

The advantage to this type of unit is that it is less expensive than the electronic version, and if the homeowner should decide to upgrade to an electronic unit, all that needs to be done is to wire the unit and replace the media filters with electronic units. In this way, you are giving the homeowner a choice. Figure 11.7 shows one of these units. Figure 11.8 shows the filter media that is used in these units.

Another version of this same type of filter system uses a stronger filtering medium and is installed in the same way as the unit I just described. This unit also can be upgraded to an electronic unit by wiring the unit and replacing the filter medium with electronic packs. Figure 11.9 shows one of these units.

Another option for the homeowner is to install a return air filter. This would be done for homes that have a single return grill such as you might find in homes equipped with electric heating systems or where a forced air heating system is located on the main floor. These filters mount in the return air grill and work in the same way as standard unit filters. Figure 11.10 shows this type of filter.

FIGURE 11.7

Media air cleaner. (*Courtesy of Honeywell, Inc.*)

FIGURE 11.8

Filter medium used in a media air cleaner. (*Courtesy of Honeywell, Inc.*)

FIGURE 11.9

Expandapac media air cleaner. (*Courtesy of Honeywell, Inc.*)

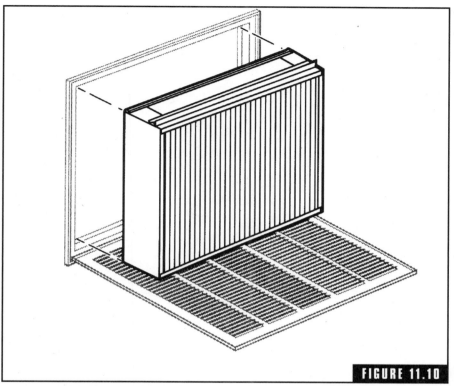

FIGURE 11.10

Return air grill Expandapac filter. (*Courtesy of Honeywell, Inc.*)

Introduction to Oil Forced Air Heating Systems

Oil forced air heating systems are found most commonly in rural settings where natural gas is not available. These types of systems come in several styles:

1. Upflow
2. Downflow
3. "Low boy"
4. Horizontal
5. Oil-wood
6. Oil-coal

This book will not provide indepth discussions of the combination heating systems because they are not very common, but I wanted to let you know that they do exist. These types of forced air heating systems work by allowing the homeowner to use an alternate type of fuel as either the primary heat source or as a backup. If the wood or coal is used as the primary heat source, the oil heating part of the unit will be used as the backup heat source. For example, if wood is used as the primary source of heat, when the wood burns down to the point where there is not enough heat being produced, the oil heating side will kick in to supply the heat. These types of combination heating systems are very expensive to have installed, but they are a great alternative to having a wood stove and an oil heating system in the home, since the wood heat will be forced into all the rooms of the home that the heating system is connected to.

As a heating professional, you will encounter the standard oil heating system more than the combination system. The oil burner will be the same in the combination system and the standard oil heating system. Both will require that you have a high level of understanding of how the system operates.

Oil heating systems require more maintenance than gas heating systems. There are more moving parts and more adjustments that are required. But an oil heating system can be one of the most energy efficient units to operate.

The oil that is burned in a home heating system is no. 2 fuel oil. This is the most refined oil that is available, and it burns very cleanly if the heating system is maintained properly. Oil burns at a much higher temperature than gas or electric heat. This means that you get more British thermal units (Btus) of heat from one gallon of heating oil than you do from a cubic foot of natural gas or a kilowatt of electricity. This makes oil heating a very good value. Here are some data that explain this last statement:

- Heating oil contains 138,690 Btus per gallon.

- Natural gas has 100,000 Btus per therm. It takes 1.4 therms to equal the heat content of 1 gallon of heating oil.

- Propane has 91,500 Btus per gallon. It takes 1.52 gallons of propane to equal the heat of 1 gallon of heating oil.

- Electricity has 3413 Btus per kilowatthour (kWh). It takes 40.6 kWh to equal the heat content of 1 gallon of heating oil.

These numbers should only be used as a comparison. Prices are different depending on which part of the country you live in, but I think you can see that oil has a higher heat output per gallon than the other types of heating fuels.

The other factor to consider is that an oil heating system is safer to operate than natural or propane gas systems. Oil has a much higher flash point (the point that the fuel will begin to burn) than gas. Fuel oil must be atomized to burn properly. This is to say that you must break the raw fuel into a mist. This is what happens in the heating cycle prior to the oil igniting.

Oil heating systems are installed in the same manner as the other heating systems in this book, with the exception that they are connected to a tank that is the source of fuel for the heating system. Some of the special precautions that must be taken with these tanks will be covered in Chap. 15.

In the next chapter I will explain the electric circuits of the oil forced air heating system.

Electric Circuits for Oil Forced Air Heating Systems

The electric circuits that control the oil forced air heating system are in some ways the same as the circuits that control the gas forced air heating system and, for that matter, most other forced air heating systems. All these systems use a series of "switches" to control the flow of fuel to the heating system and to control when and how long the system operates. These circuits use either low voltage (24 V ac), high voltage (110 V ac), or in the case of the oil heating system, ultrahigh voltage (14,000 V ac). This ultrahigh-voltage circuit is used for ignition of the fuel oil.

Low-voltage circuits primarily control the thermostat and relay circuits. A step-down transformer is used to convert the 110-V ac voltage to the 24-V ac output voltage. In some cases, the thermostat may run directly off the 110-V circuit. You can tell this by looking at the size of the wire that is used. On a 24-V circuit, the wire is very thin and normally will be two colors, red and black. It also will have a single copper wire. On the 110-V circuit, the wire will be much larger in diameter, and the color may be black and white. Of course, you should always check the voltage used with a voltmeter to make sure of the voltage you are working with.

Let's begin by looking at the low-voltage circuit on the oil forced air heating system. On such an oil forced air heating system, one of two types of primary controls will be used. The system will either have a stack relay control or a cad-cell relay control. In either case, this is the control that contains the step-down transformer that controls the low-voltage circuit. On some of the older models, this transformer may be separate from the relay controls. In this case, the transformer may be mounted in a wire cabinet or mounted to the outside of the unit.

This same low-voltage circuit is used to control the relays that operate the combustion blower, cad cell (when used), and blower relay. The thermostat is wired to the transformer to act as a switch. One wire runs from the transformer to one side of the thermostat. The other wire runs from the other terminal of the thermostat to one side of the combustion blower relay. The other terminal from the transformer is connected to the other side of the combustion blower relay. Thus, when the thermostat calls for heat, this circuit is complete, and the combustion blower begins to operate. This, in turn, starts the oil pump turning by means of a coupling between the combustion air motor, which supplies oil to the gun assembly.

At this point, the cad-cell relay is energized. And once the unit reaches the proper operating temperature, the blower relay is energized, and the blower begins to operate in the heating mode.

On the high-voltage side, the 110 V is supplied directly from the fuse panel or breaker box of the home. This power normally is connected to a switch at the heating unit that has a fuse attached. This switch, also known as an *SPST switch,* is located on the outside of the unit.

When we are talking about an oil heating system, the 110-V power is connected to the stack control or cad-cell relay. All the other major 110-V circuits are connected to one of these devices by means of a series of relays.

As the heating cycle begins and the relays are energized by the low-voltage circuits, 110-V power is supplied to the combustion air blower motor. This motor is connected to an oil pump by means of a coupling.

Once the combustion blower motor is running, the oil pump begins to supply fuel oil to the gun assembly. At this point, the ignition circuit is energized, and 110 V is supplied to the ignition transformer.

The ignition transformer converts the 110-V power to ultrahigh voltage by means of a step-up transformer. This step-up transformer converts the 110-V power to 14,000 V ac. It is this voltage that is used to ignite the atomized fuel to begin the heating cycle.

Once the ignition sequence has begun, the cad-cell relay (which is used to sense the light of the fire burning) or on older systems the stack control (which is used to sense the heat that is generated by the ignition sequence) begins to time the system. If the cad-cell sensor or stack control does not sense the light or heat within 45 s, it will shut the unit down. This is one of the main safety features that is designed into the system. Once the relay "times out," the reset switch must be pushed to attempt to restart the heating cycle. As you will learn in Chaps. 17 and 18, you must assume that the homeowner has attempted to restart the heating unit several times before you are called to service the unit. Unfortunately, the homeowner frequently is not aware that by overriding this safety, he or she is allowing the unit to pump raw fuel oil into the fire pot. As you will learn in Chap. 18, there is a proper way to burn off this raw fuel and the dangers that it presents.

Once the heating cycle has begun, the fan control will act as another form of switch to control the 110 V to the blower motor. Once the temperature reaches the set point of the fan control, power will be sent to the blower motor, and the forced air part of the heating cycle will begin. If the blower does not come on for any reason, the fan control will shut the power off to the unit so that it does not overheat. This is another safety feature of the unit.

By understanding how these circuits operate, you will be able to isolate the problem when you are called to troubleshoot an oil forced air heating system. By understanding how the low-voltage, high-voltage, and ultrahigh-voltage circuits operate, you will have a better understanding of how one circuit operates in conjunction with the other circuits. This, of course, has been a general overview of the circuits. Figure 13.1 shows how all the circuits of an oil forced air heating system are connected. Study this wiring diagram so that you will be better prepared when the time comes to troubleshoot one of these systems.

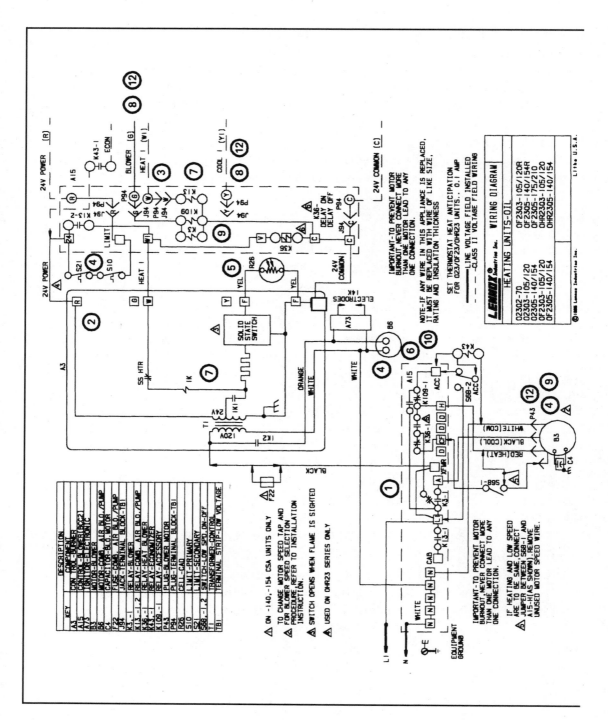

140

O23 / OHR23 / OF23 OPERATION SEQUENCE:

1- When disconnect is closed, 120V is routed to the blower control board BCC2 (A15). The BCC2 feeds line voltage to transformer (T1).

2- T1 supplies 24VAC to the burner control (A3). In turn, A3 supplies 24VAC to A15, which supplies the indoor thermostat (not shown) with 24VAC.

3- When there is a call for heat, W1 of the thermostat energizes W of the A15 board with 24VAC.

4- A15 energizes combustion air blower relay (K13). When K13-2 closes, assuming primary limit (S10) and [secondary limit (S21) in OHR units only] are closed, 24VAC is routed to 1K, which closes 1K1 and 1K2. When 1K2 closes combustion air blower / pump (B6) is energized. Simultaneously 24VAC is routed through the blower relay (K36). When K36-1 closes blower motor (B6) is energized on heating speed.

5- When 1K2 closes the electronic ignitor is energized causing a 14,000VAC spark, igniting the burner. When 1K1 closes the solid state switch and cad cell are energized. When the cad cell senses light the solid state switch de-energizes the safety heater, keeping the burner operating.

6- A15 energizes accessory relay (K109). When K109-1 closes the accessory terminal on the A15 board is energized with 120VAC.

7- When heat demand is satisfied, W1 of the thermostat de-energizes W of the ignition control. W de-energizes K13 in turn de-energizing 1K. When 1K is de-energized B6 and A73 stop immediately. The indoor blower runs for a designated fan "off" period (90–330 seconds) as set by jumper on BCC2 control.

8- When there is a call for cooling, Y1 of the thermostat energizes Y and G of the A15 board with 24VAC.

9- A15 energizes blower relay K3. When K3-1 closes B3 starts on the speed set for cooling.

10- A15 energizes accessory relay K109. When K109-1 closes the accessory terminal on the A15 board is energized with 120VAC.

11- Provided that condensing unit is connected to Y terminal, cooling will start.

12- When cooling demand is satisfied, Y1 of the thermostat de-energizes Y and G. The indoor blower and condensing unit stops immediately.

FIGURE 13.1

Wiring diagram for an oil forced air heating system. (*Courtesy of Lennox Industries, Inc.*)

Controls for an Oil Forced Air Heating System

Now that we have been introduced to oil forced air heating systems and have examined their electric circuits, let's take a look at the components of the system. Figures 14.1 through 14.3 show, three different types of oil forced air heating systems with their respective parts layout. As you can see from these illustrations, the oil forced air heating system requires more maintenance than the gas forced air heating system. There are more parts to contend with, and the settings of these controls are more critical for proper operation of the system. It is also extremely important that the settings for the electrodes are accurate, or you can have a delayed ignition problem and much lower efficiency from the unit. All these items and more will be covered in the chapters that follow on tuneup and troubleshooting oil forced air heating systems (Chaps. 17 and 18). For now, let's look at the controls for oil forced air heating systems.

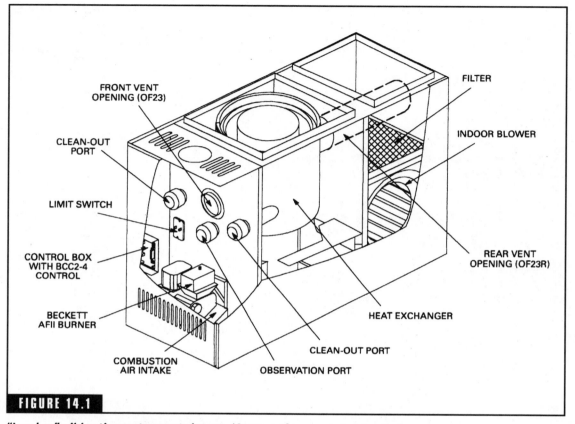

FRONT VENT OPENING (OF23)

CLEAN-OUT PORT

LIMIT SWITCH

CONTROL BOX WITH BCC2-4 CONTROL

BECKETT AFII BURNER

COMBUSTION AIR INTAKE

OBSERVATION PORT

CLEAN-OUT PORT

HEAT EXCHANGER

REAR VENT OPENING (OF23R)

INDOOR BLOWER

FILTER

FIGURE 14.1

"Low boy" oil heating system parts layout. *(Courtesy of Lennox Industries, Inc.)*

1. *Oil tank.* This is the unit that contains the fuel oil that is burned in the oil forced air heating system. This tank can be placed in several locations:

   a. Outside the home

   b. Inside the home, in the basement or near the heating unit

   c. *Underground.* In this type of arrangement, you must be extremely careful that there are no leaks from the tank that can contaminate the groundwater source. If you are using one of these tanks and it is over 10 years old, you should have it checked out by your local oil company. Figures 14.4 and 14.5 show the locations of underground and inside tanks. Chapter 15 will go into more detail on how to protect the oil tank during the winter months so that you can help to avoid problems with the heating system in the winter.

2. *Burner assembly.* The burner assembly is the main combustion unit of the oil forced air heating system. Figures 14.6 and 14.7 show the burner assembly.

3. This unit contains several components. They are

    a. *Oil pump motor.* The oil pump motor is mounted on the side of the burner assembly. The main function of the oil pump motor is to drive the oil pump and induction blower. They are connected by a burner coupling.

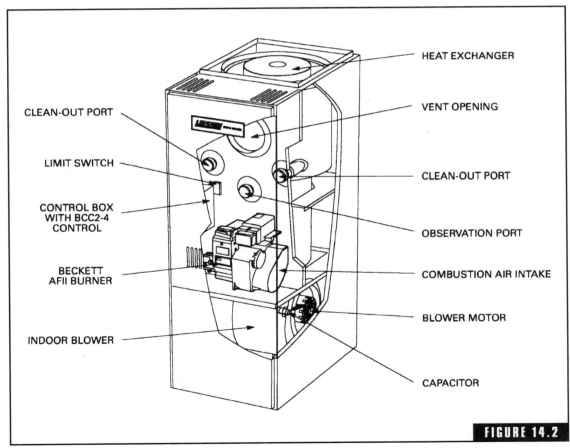

FIGURE 14.2

Upflow oil forced air heating system. (*Courtesy of Lennox Industries, Inc.*)

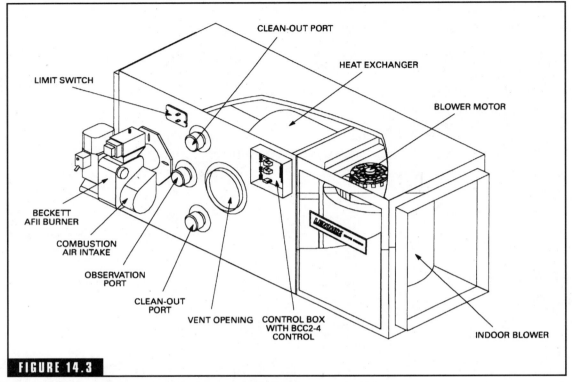

FIGURE 14.3

Horizontal oil forced air heating system. (*Courtesy of Lennox Industries, Inc.*)

b. *Burner coupling.* This is the connection between the burner motor and the oil pump to act as the drive shaft for the oil pump. The oil pump generates between 100 and 140 psi to move oil through the gun assembly.

c. *Oil pump.* The oil pump is used to deliver oil from the oil storage tank to the gun assembly. The pump normally is located on the left side of the burner assembly. It is connected to the burner motor by the burner coupling. This pump delivers oil to the nozzle at pressures of 100 to 140 psi. The reason for the high pressure is so that the oil can be atomized by the nozzle. Figure 14.8 shows an oil pump.

d. *Blower wheel.* The blower wheel is located between the burner motor and the oil pump. The purpose of the blower wheel is to introduce air into the combustion cycle that occurs when the oil is ignited. This wheel typically is attached to the burner motor shaft. An air adjustment gauge is located on the blower wheel housing and is used to adjust the amount of combustion air for proper ignition. Figure 14.7 shows the location of this wheel.

e. *Ignition transformer.* This device is used to deliver ultrahigh voltage to the electrodes to ignite the atomized fuel at the nozzle. This transformer delivers 14,000 V ac to the electrodes. It is located on the burner assembly and attaches to the electrodes either by direct contact with the electrode transfer bars or by wires. Figure 14.9 shows an ignition transformer.

f. *Gun assembly.* The gun assembly contains the electrodes and nozzle for the ignition cycle. The main purpose of the gun assembly is to transfer oil from the oil pump to the nozzle. The gun assembly is connected directly to the oil pump. The gun assembly is located inside the burner assembly and is accessed either by opening the access door or by removing the ignition transformer. Figure 14.10 shows a gun assembly.

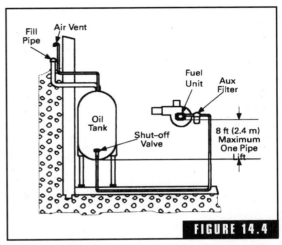

FIGURE 14.4

Inside of the home oil tank. (*Courtesy of Lennox Industries, Inc.*)

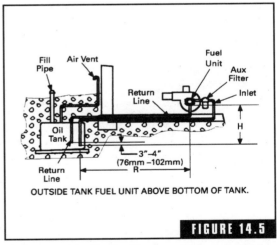

OUTSIDE TANK FUEL UNIT ABOVE BOTTOM OF TANK.

FIGURE 14.5

Underground fuel tank. (*Courtesy of Lennox Industries, Inc.*)

g. *Nozzle.* The nozzle is used to convert raw heating oil into a form that can be burned in the combustion process. This nozzle filters any particles that do get past the oil filter. It then forces the oil that is delivered under pressure from the oil pump through a fine opening in the end of the nozzle that atomizes the fuel. Nozzles come

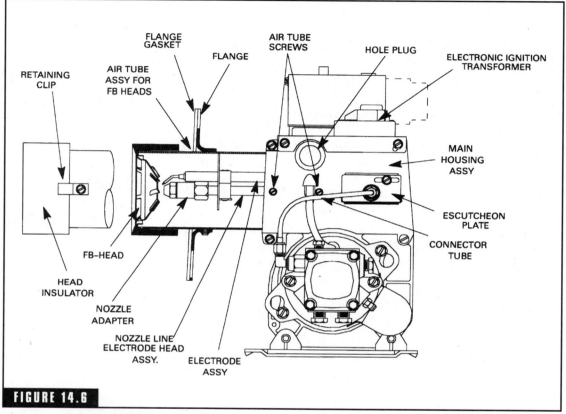

FIGURE 14.6

Side view of burner assembly. (*Courtesy of Lennox Industries, Inc.*)

in many different sizes. The size of the nozzle is normally referred to in its gallons per hour (gph) rating. This means that a nozzle with a rating of 1.0 gph would burn 1 gallon of fuel per hour. The other rating on the nozzle refers to the angle of the spray. This is very important because the head is designed to accommodate this angle. The rating of a nozzle would be expressed, for example, as 1.0 × 80. This would mean that the nozzle would burn 1 gallon of fuel per hour and would have a spray pattern of 80 degrees. Always replace the nozzle with the same gallon and spray pattern rating. In some cases, you may want to change the gallon rating, but do not change the spray rating. Nozzles also come in four different types. They are hollow, semisolid, solid, and extrasolid. Solid nozzles are used on long chambers, whereas a shorter chamber might use a hollow or semisolid nozzle.

h. *Electrodes.* The electrodes are used to ignite the fuel that is atomized in the nozzle. They are attached to the gun assembly and the ignition transformer as described earlier. The electrode spark gap is one of the most critical settings on the oil forced air heating system. You must make sure that the gap between the electrodes is

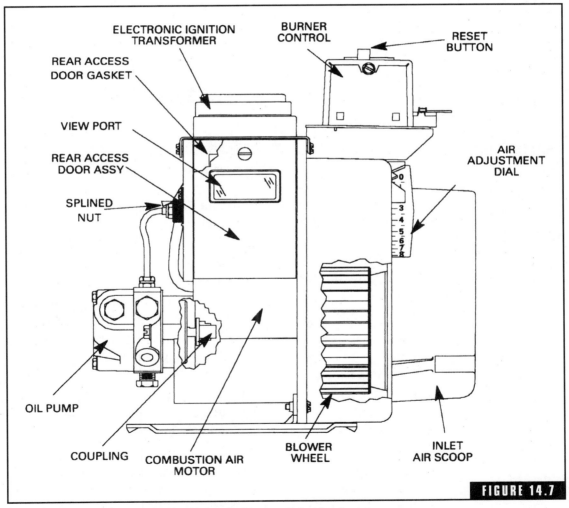

FIGURE 14.7

Front view of the burner assembly. (*Courtesy of Lennox Industries, Inc.*)

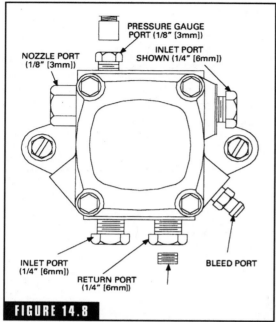

PRESSURE GAUGE
PORT (1/8" [3mm])

INLET PORT
SHOWN (1/4" [6mm])

NOZZLE PORT
(1/8" [3mm])

INLET PORT
(1/4" [6mm])

RETURN PORT
(1/4" [6mm])

BLEED PORT

FIGURE 14.8

Oil pump. (*Courtesy of Lennox Industries, Inc.*)

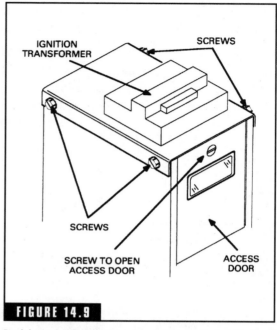

IGNITION
TRANSFORMER

SCREWS

SCREWS

SCREW TO OPEN
ACCESS DOOR

ACCESS
DOOR

FIGURE 14.9

Ignition transformer. (*Courtesy of Lennox Industries, Inc.*)

correct. You also must ensure that the distance between the electrode and the center of the nozzle is correct. In addition, the distance from the center of the nozzle and the end of the burner flame retention head must be correct. Figure 14.11 illustrates how these settings are done with the use of an AFII gauge.

i. Cad cell. The cad cell is one of the main safety features on an oil forced air heating unit. The job of the cad cell is to prove the flame from the heating cycle. This cell works with the burner control to allow the heating cycle to continue. If the cad cell cannot prove the flame, it will open the circuit to the burner control and shut the power off to the burner motor and ignition transformer. The cad cell is located inside the burner assembly behind the access panel. Figure 14.12 shows this location.

j. Burner control. This is the device that controls the operation of the burner assembly. It is used with the cad cell to allow operation of the burner assembly once the flame from the ignition process can be proved. If the flame cannot be proven, then burner control will shut the power off to the burner assembly. There is a manual reset button on the top of the control that must be reset once the cause of the problem has been resolved. As you will discover in Chaps. 17 and 18, this is one of the primary control devices for the oil forced air heating system. It is also the device that the home-owner will use to attempt to get the heating system to operate in the event of a failure. The problem with doing this is that if the flame cannot be proven, all that is happening is that raw fuel is being pumped into the fire pot. This fuel will then have to be burned off once the repair has been made and before the unit can be put back into normal operation. This can be a major problem for the heating professional. Figure 14.6 shows the location of this device.

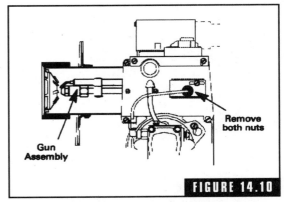

Gun assembly. (*Courtesy of Lennox Industries, Inc.*)

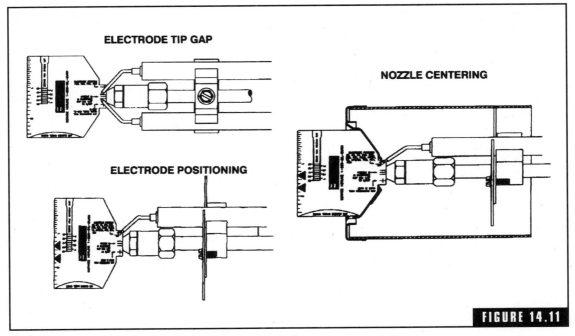

Electrode gap setting. (*Courtesy of Lennox Industries, Inc.*)

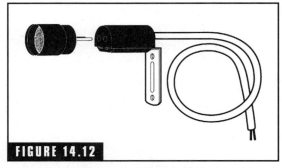

FIGURE 14.12

Cad-cell relay. (*Courtesy of Honeywell, Inc.*)

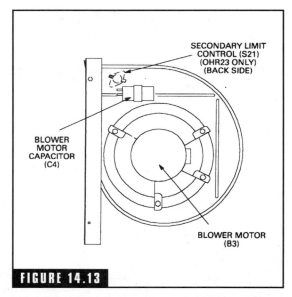

SECONDARY LIMIT
CONTROL (S21)
(OHR23 ONLY)
(BACK SIDE)

BLOWER
MOTOR
CAPACITOR
(C4)

BLOWER MOTOR
(B3)

FIGURE 14.13

Side view of an indoor blower. (*Courtesy of Lennox Industries, Inc.*)

k. *Oil filter.* The oil filter is used to remove any particles of dirt or debris that may be in the oil from the oil tank. Most oil forced air heating systems use no. 2 fuel oil, which is the cleanest fuel oil available. However, foreign matter can still be present in the tank, and this can cause a problem with the heating system if the oil is not filtered properly. The oil filter can be located at the oil tank, or it can be located just before the oil pump. In either case, it is important to change this filter on a regular basis for proper operation of the heating system. Care must be taken to ensure that once the filter is changed, all the air is bled from the system. This will be covered in more detail in Chaps. 17 and 18.

l. *Limit switch.* The limit switch is used to control the unit in the event of an indoor blower failure. Figure 14.2 shows this location. This safety device will open the circuit and shut down the burner assembly so that the unit will not overheat.

m. *Indoor blower assembly.* The indoor blower is used to blow heat created from the combustion process from the burner assembly into the home. The indoor blower assembly consists of the blower motor, a blower cage that is attached to the blower motor, and a capacitor that is used to help start the blower motor. Figure 14.13 shows the indoor blower assembly. The indoor blower is controlled by the blower control board. This board controls the time interval between the start of the burner assembly and when the indoor blower begins to blow the warm air into the home. This board also serves many other functions in an oil forced air heating system and can be seen in Fig. 14.14.

n. *Heat exchanger.* The heat exchanger is used to observe the heat that is generated from the combustion process. When the indoor blower begins its cycle, it blows air around the heat exchanger to warm the home. The heat exchanger is also used to transfer the fumes that are a by-product of the combustion process to the chimney to vent them to the outside of the home. This is one important factor to be considered when you read Chap. 17. It will become clear why it is important to perform a check of the heat exchange during this process. If there is any damage to the heat exchanger, the heating unit must not be allowed to operate until either the heat exchanger is replaced or the heating system is replaced. Damage to the heat exchanger is a very dangerous situation.

This chapter has covered a lot of information, but it is important that you have a complete understanding of how the different parts of an oil forced air heating system operate. As with all heating systems, all the components must operate together for the system to function properly.

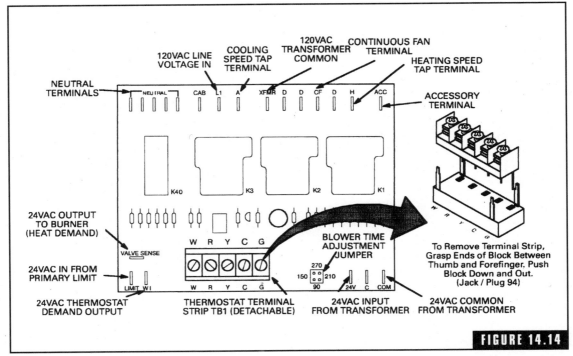

Blower control board. (*Courtesy of Lennox Industries, Inc.*)

FIGURE 14.14

Study the illustrations in this chapter so you have a good understanding of the location of each component. When you begin Chap. 17 on tuning up an oil forced air heating system, you will need this information. The next chapter diverges briefly to describe the protection of oil tanks in the winter.

Protecting Oil Tanks in the Winter

One of the most important items that you must think about in the winter is protecting your oil tank. By this I mean that you must take precautions to keep the oil from freezing. Oil tanks that are outside the home are exposed to all the elements in the winter. They must withstand the subzero temperatures, snow, ice, rain, etc. If you do not protect the tank from the elements, you may find that you will not have any heat one morning because the oil line between the tank and the house has frozen. However, there are a few simple precautions you can take to protect your family from this problem.

Heat Tape

Heat tape is an item that you wrap around the exposed oil line. The tape plugs into a normal household outlet and will keep the oil line warm. If you are going to install heat tape on your oil line, you must make sure that when you wrap the line you do not allow the tape to be in contact with itself. This means that when you wrap the oil line, you must not allow the tape to overlap onto itself. This may cause the tape to short out or have a shorter life span.

Insulation

Another way to protect your oil line is to install insulation on the line. There are a couple of different types of insulation that will work:

1. *Bat-type insulation.* This is the same type of insulation that is used in the home to insulate the walls and ceilings. This type of insulation will work, but it must be wrapped around the pipe and oil filter and secured in place with gray tape. This will insulate the pipe and oil filter, but it is not intended for use in exposed situations. You will have to replace this type of insulation each year.

2. *Pipe insulation.* This is a foam insulation that is made to insulate pipe. It comes in different sizes. One side of the insulation has a slit in it so that the insulation can be placed over the pipe. This type of insulation works extremely well and will last for several years.

Whatever type of insulation you use is your decision, and either type will work. The important thing to remember is to do something to protect the exposed line.

Before I moved my family to Oregon, we lived most of our lives in Michigan. The area that we lived in had a lot of exposed country. I can tell you that receiving a call at 3 A.M. to go to a client's home because he or she had no heat in the dead of winter was not fun. What was worse was to discover, once I arrived, that the problem was that the oil line had frozen. There was no insulation on the line, and the temperature was below zero. The only thing that you can do is to get the propane torch out of the truck and begin to get the oil line thawed. I always carried insulation in my truck so that when I was on a call and the line was exposed, I could wrap it before I left. It is always a good idea to check the line at the time you are called to do a tune-up of the system in the summer. If there is no insulation, either recommend that the client wrap the line or have him or her allow you to do the work. You will be saving both of you a lot of work and inconvenience later.

Another problem with oil tanks is that they collect water and sludge. Because an oil tank is made of metal, it will collect condensate when the temperature changes. Water is lighter than the oil in the tank, so it will collect at the top of the oil. In some cases, this water can be an inch or more deep. If the oil is allowed to drop too far in the tank, the water will begin to be drawn into the heating system. Since water does not burn, you will have no heat. I will cover this problem in more detail in the next couple of chapters, but it also needs to be addressed here.

If you have a client who cannot afford to keep the oil tank full, you may run into this problem. Every time that the tank is filled, all the water and sludge will be moved around. If the tank is kept full, this should not be a problem. However, if the tank is allowed to drop below 22 gallons, or one-eighth of a tank, you can have a potential problem. If this is the case, you need to advise the client that he or she needs to have the tank filled more often, or if the problem is severe, he or she will have to contact the oil company to have the tank pumped out to get rid of the problem. You need to understand the reading on the gauge to know how much oil is in the tank when you begin to troubleshoot the system when there is a problem. By knowing this information and asking the proper questions of the homeowner, you can save a lot of troubleshooting time by finding out that the problem is in the tank. If the tank has been allowed to drop too low and the client could not afford to put much oil in the tank, you will have to attempt to bleed the water from the line to get the unit running again. The procedure for this will be covered in detail in Chap. 18.

Another tip for the homeowner is to put an additive in the oil tank with every oil fill. One of these additives is called "dry gas." This additive comes in a small plastic bottle and is available in most automotive stores and service stations. It is added to the oil tank after an oil fill and will help to keep moisture from forming. It will burn along with the fuel oil and will not harm the system.

I have discussed several things for the homeowner to do to protect the oil tank in the winter. The cost of this type of prevention is very small. In some cases, the homeowner may have the items needed in the home already. It is very difficult to get some homeowners to understand the need to protect the oil tank and line in the summer. They do not want to think about the problems of the winter while the weather is so nice. It is part of your job as a heating professional to convince the homeowner that this is the right thing to do. You might even suggest to the owner of your shop to allow the service group to have some insulation on the truck and to offer this as a free service to the client. This could be a great PR item for your shop.

The next chapter addresses the operation of an oil forced air heating system. You will begin to see how all the items in the preceding chapters on oil forced air heating systems come together to make the system work.

Operation of an Oil Forced Air Heating System

The operation of an oil forced air heating system has not changed much over the course of time. By this I mean that while there have been great strides in the efficiency of these units, the basic operation has remained the same.

Chapter 5 discussed the operation of a gas forced air heating system. In that chapter I mentioned that there are new electronic ignition systems and the old standing pilot systems. This is not the case with oil forced air heating systems. Most of the advances have been made in the controls side of the unit. Stack controls have been replaced with cad-cell relays. Wire connection boxes have been replaced by integrated controls, but the basic operation of the heating system has remained the same.

This chapter is devoted to explaining to you how an oil forced air heating system works and how the different controls and circuits that you studied in the preceding chapters come together to make the system work. You should begin to see how the subjects covered in Chaps. 12 through 15 work together to make the system work.

Sequence of Operation

1. The *thermostat* (Fig. 16.1) is either turned up or begins to call for heat. The circuit is closed and sends a signal to the controlling relay (Fig. 16.2).

2. The *control relay* (this can be a stack control or a cad-cell control) then closes the circuit and sends power to the oil pump motor (Fig. 16.3) and ignition transformer (Fig. 16.4) to begin the heating cycle.

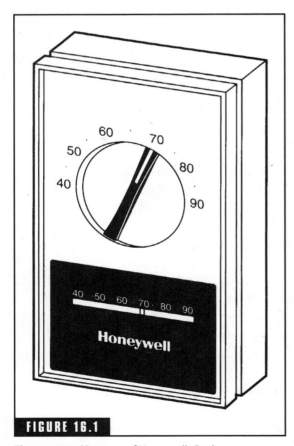

FIGURE 16.1

Thermostat. (*Courtesy of Honeywell, Inc.*)

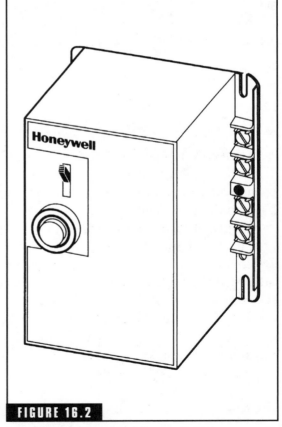

FIGURE 16.2

Control relay. (*Courtesy of Honeywell, Inc.*)

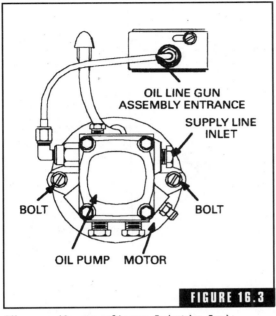

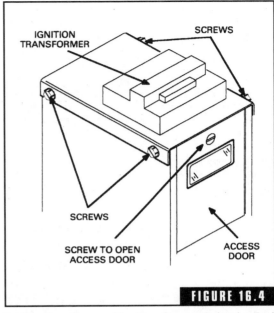

FIGURE 16.3

Oil pump. (*Courtesy of Lennox Industries, Inc.*)

FIGURE 16.4

Ignition transformer. (*Courtesy of Lennox Industries, Inc.*)

3. The *oil pump motor,* connected by a coupler, begins to turn the oil pump. The oil pump begins to draw fuel oil from the oil tank.

4. The *ignition transformer* is energized and begins to spark between the points on the electrodes.

5. As the oil begins to flow from the oil pump down the *gun assembly* (Fig. 16.5), it is atomized in the nozzle.

6. The fuel oil that is atomized by the *nozzle* is then ignited by the *electrodes,* and the heating cycle begins.

7. Once the cycle has begun, the timer in the cad-cell relay (Fig. 16.2) is activated. If the *cad cell* (Fig. 16.6) does not "see" the light from the flame, the circuit is opened, and the cycle is stopped. This is a safety feature of the system so that raw fuel oil is not pumped into the fire pot.

8. If the oil forced air heating system does not use a cad-cell relay, it will have a *stack control* (Fig. 16.7). This control works like the cad cell except that it "looks" for the heat that is generated from the ignition cycle to prove that the system is operating. If the stack relay

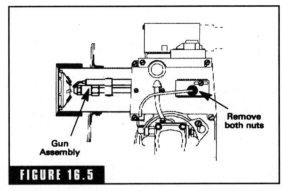

FIGURE 16.5

Gun assembly.

Remove both nuts

Gun Assembly

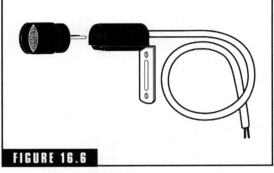

FIGURE 16.6

Gun assembly. (*Courtesy of Lennox Industries, Inc.*)   **Cad cell.** (*Courtesy of Honeywell, Inc.*)

cannot prove the heat, it will open the circuit and stop the cycle. In both these cases, the unit must be reset by pressing the *reset button* on either the cad-cell relay or the stack control. Caution must be exercised when pressing the reset button to make sure that the system will ignite. Each time that the reset button is pressed and the system does not ignite, you are allowing raw fuel to enter the oil pot. When the system is repaired, this raw fuel will need to be ignited and allowed to burn off. This will be covered in Chap. 18.

9. Once the system is operating and the flame or heat has been proven, the system will continue to heat until the temperature is high enough to close the circuit on the *fan control* (Fig. 16.8).

10. The fan control is the switch that turns the indoor blower on and off. Once the temperature has reached the on set point, the indoor blower will begin to blow forcing warm air into the home.

11. Once the temperature in the home has reached the setting on the *thermostat,* the circuit will open and the burner assembly (Fig. 16.9) will shut down.

12. The indoor blower will continue to blow until the temperature in the *heat exchanger* reaches the lower setting on the fan control and the circuit is opened, stopping the indoor blower.

13. One point that I would like to make is that the fan control is also a safety. In the event of an indoor blower motor failure, once the temperature reaches the high limit on the fan control, the circuit will be opened and the unit will shut down. This may sound con-

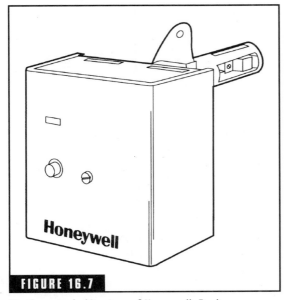

FIGURE 16.7

Stack control. (*Courtesy of Honeywell, Inc.*)

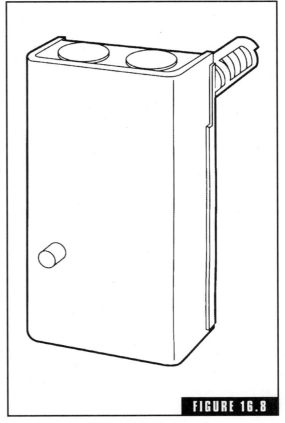

FIGURE 16.8

Fan control. (*Courtesy of Honeywell, Inc.*)

fusing to some of you since I stated earlier that the indoor blower circuits are independent of the rest of the system. This is the one exception to this rule. For most troubleshooting problems that deal with the indoor blower, the blower will not be connected with the rest of the system. The exception is the high limit. I hope that this does not confuse you, but all the safeties are connected together so that if there is a system failure, the whole system will shut down.

Once you read the next two chapters on tune-up and troubleshooting, this point should become more clear. All modern heating systems have this type of safety feature built in to protect both the equipment and the homeowner.

We have covered all the basics of an oil forced air heating system. It is now time to explore how all the controls and circuits that you have read about in the preceding chapters come together to make the oil forced air heating system work.

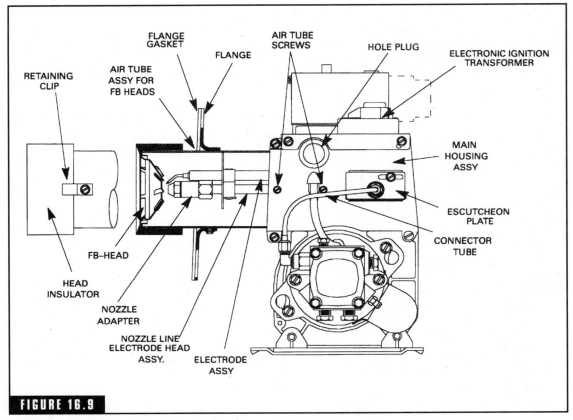

RETAINING CLIP

AIR TUBE ASSY FOR FB HEADS

FLANGE GASKET

FLANGE

AIR TUBE SCREWS

HOLE PLUG

ELECTRONIC IGNITION TRANSFORMER

MAIN HOUSING ASSY

ESCUTCHEON PLATE

CONNECTOR TUBE

FB–HEAD

HEAD INSULATOR

NOZZLE ADAPTER

NOZZLE LINE ELECTRODE HEAD ASSY.

ELECTRODE ASSY

FIGURE 16.9

Burner assembly. (*Courtesy of Lennox Industries, Inc.*)

The next chapter on tune-up will teach you how to check the different circuits and systems in the heating unit to make sure the system will perform properly during the heating season. I also will explain how to replace faulty components that you will encounter in real life when you do this type of maintenance on the equipment. The knowledge that you gain from these chapters will prepare you for the chapter on troubleshooting. Of all the forced air heating systems covered in this book, oil forced air heating systems require the most amount of maintenance. It is also the most fascinating system to work on. There will be more interaction between you as a heating professional and the system to make it operate properly. Since this type of system has more operational parts than gas or, as you will learn, electric forced air heating systems, I will spend much more time explaining these components in the next two chapters.

I will include as many drawings and charts as I can in this next chapter so that you get a clear understanding of the subjects covered. There are many items that will be critical to proper operation of the system, such as the gap of the electrodes and the voltage that is put out by the ignition transformer.

There are many more chances for failure in a oil forced air heating system than in the other system, but the cause of the failure can be traced to the source quickly if you know what to look for, and that is what you will learn in the next two chapters. If there were any items in the preceding chapters on the oil forced air heating system that you did not understand, take the time to go back over the section before reading on. If you feel that you have a good understanding of the subjects covered, then move on to the next chapter.

Tuning Up an Oil Forced Air Heating System

Of all the forced air heating systems in this book, the oil forced air heating system is the most challenging. There are many more parts to examine and many more parts that must be replaced during the tune-up. Properly tuned, an oil forced air heating system is one of the most economical heating systems available.

Fuel oil has one of the highest Btu ratings of any fuel that is used in the heating systems described in this book. Because of this fact, it is critical that you know how to size the nozzle to the heating system that you are working on. I will give you the conversion calculation that will show you which nozzle size to use for the Btu rating of the forced air heating system that you are tuning up. A complete chart is included in Appendix C.

Before we begin the tune-up of an oil forced air heating system, there are certain tools that you will need to make the job go quickly. An experienced heating professional can tune up an oil forced air heating system in about 1½ hours. This time can change depending on the condition of the heating system and the number of parts that need to be replaced. The basic list of tools is given below.

Let's begin by replacing the oil filter, examining the condition of the oil tank, examining the oil burner assembly, and finishing with the blower assembly. One thing that you will notice is that blower assembly maintenance and repair are the same as for a gas forced air heating system. The operation of this device does not change with the change in fuels used. Let's begin with replacement of the oil filter and tank examination.

Oil Tank

The first thing that you want to do before starting any repair or maintenance on a heating system is to turn the power off to the unit. Locate the switch that controls the power to the unit. If you remember from the preceding chapters, this will be located on either the

---

TOOLS

THIS IS ONLY A BASIC LIST OF THE TOOLS NEEDED TO PERFORM THE SUMMER TUNE-UP OF AN OIL FORCED AIR HEATING SYSTEM, BUT YOU CAN SEE THAT IT IS MUCH MORE EXTENSIVE THAN THAT FOR A GAS FORCED AIR HEATING SYSTEM.

1. Drop light or flashlight
2. Open-end wrenches
3. Box-end wrenches
4. Adjustable wrench
5. Allen wrenches
6. Oil can with 10W40 oil
7. Catch pan
8. Electrical tester
9. Transformer tester for the ignition transformer
10. Pressure gauge
11. Dry gas
12. Hammer
13. Oil sorb
14. Industrial vacuum cleaner
15. Inspection mirror
16. Shop rags
17. Thermostat wrench
18. Bearing grease
19. Nozzle wrench (two adjustable wrenches can be used)
20. Metal file
21. Tape measure
22. Electrode gauge (optional)

heating unit or in the fuse box or breaker panel. Once you have located the power, turn it off, and check that it is off by operating the blower circuit manually. As you will recall, this can be done at either the thermostat or on the blower control on the heating system. Once you are sure that the power is turned off, you can begin.

Locate the oil tank for the unit that you are working on. The oil tank can be in one of three locations:

1. Outside the home above ground

2. Outside the home below ground

3. Inside the home close to the heating unit

Figures 17.1 and 17.2 show two of these possible tank and filter locations.

Once you have determined the location of the oil tank, locate the shut-off valve. On tanks that are located above ground, the shutoff will, in most cases, be located at the tank. If the tank is located below ground, the shutoff can be located either at the line as it enters the home or just before to the burner assembly. Turn off the valve

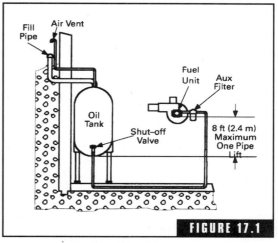

Above-ground tank and piping.

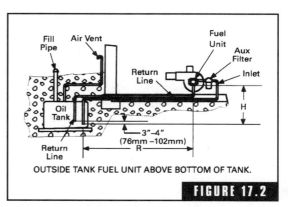

Two-pipe below-ground oil tank.

that supplies the fuel from the tank. Now find the filter. This will be located close to the shutoff valve. The filter is located in a metal canister and is inline with the oil feed line. Locate the filter now.

Once you have located the filter, you are going to replace it with a new filter cartridge. This must be done every year to keep the oil forced air heating system operating at maximum efficiency. Place your catch pan under the canister. There is a bolt located in the center of the top of the canister. You will need to loosen this bolt until the canister bottom can be removed. This is where the filter is located.

Once you have the canister bottom removed from the top, dump the oil out of the canister, and remove the oil filter. There is also a seal that is located around the lip of the canister. In some cases, if the filter has not been changed for a long period of time, the seal can be "stuck" to the lid of the canister. In either case, remove the old seal because you will need to replace this as well.

Replace the filter cartridge, and place the seal in the groove on the lip of the canister. Position the bottom of the canister in place with the top of the canister, and tighten the bolt so that the canister bottom and top are tight and sealed. Do not overtighten the bolt. The bolt must be tight enough to keep air from entering the system and to keep oil from leaking out as well.

Since the oil supply is a "sealed" system, meaning that there can be no air in the line, you must "bleed" the air from the system. Anytime that you "break" the oil line, you run the risk of having air enter the system. This can cause the oil burner to sputter. It also can cause the oil pump to "vapor lock" in some cases, and this means that the oil pump cannot get the suction needed to pull oil from the tank. This will cause the oil forced air heating system go into a reset mode and not operate. Therefore, you can see how important it is to get all the air out of the system. You may run in to these types of situations when you have to troubleshoot this type of forced air heating system, and you now know the cause. We are getting ahead of ourselves, however, so let's get back on track.

Located on the top of the canister is a smaller bolt. This is the bleeder bolt. Find the wrench that fits this bolt before we go on. Once you have the right wrench, loosen this bolt slightly. Open the oil shut-off valve from the tank to allow oil to flow into the oil canister. You will need to adjust the bleeder bolt to allow for the air to escape from the system. The best way to do this is to wiggle the bolt. You should hear the air escaping and possibly see bubbles in the air. Continue to bleed the system until you have a good flow of oil from the bleeder bolt. Once you have determined that you have all the air out of the system, close the bleeder bolt and tighten it. This has now removed all the air from the system so that there is a good oil flow to the burner assembly.

Another item that will cause air to enter the system is if the homeowner has allowed the oil tank to run out. As long as there is oil in the tank, air cannot enter the system from this point. You should let the homeowner know that he or she should keep a minimum of one-quarter tank of oil at all times. This will keep air from entering into the system and possibly causing a service call in the winter.

Before you leave the oil tank, you should dump a bottle of dry gas into the tank. This will keep moisture from forming in the tank during the summer and will help to keep the oil line from freezing in the winter. You should advise your client to put a bottle of dry gas into the tank several times in the winter because this is an inexpensive way to avoid winter problems.

If the tank is located outside the home above ground, you also should look at the condition of the oil supply lines. Is there heat tape or insulation on the line? If there is not, you should advise the client that this is an item that needs to be addressed. If the home is located in the country or is exposed to the north wind, the chance that this line will freeze and cause the heating system to fail in the winter is quite high. You should offer the service of installing heat tape or insulation as part of the tune-up process. Chapter 15 goes into more detail on how to perform this service.

With all this finished, the oil tank has been serviced. We will now move on to burner assembly service and operation.

Burner Assembly

The burner assembly is one of the most critical components of an oil forced air heating system. This unit houses the oil pump and motor, ignition transformer, cad cell, gun assembly, oil nozzle, and electrodes. Figures 17.3 through 17.5 show the location of these and all the components of an oil forced air heating system. Figures 17.6 and 17.7 show the components of the burner assembly that you will be servicing in this chapter.

Before attempting any maintenance or repair on a heating system, you must make sure that the power is turned off to the system. I know that I keep bringing up this point, but you can sustain serious injury if the power is not turned off to the unit prior to beginning this type of operation. It is better to check this item twice than to think that it is off only to find out as you are doing the maintenance that the motor begins to turn. This is not a pleasant position in which to find yourself.

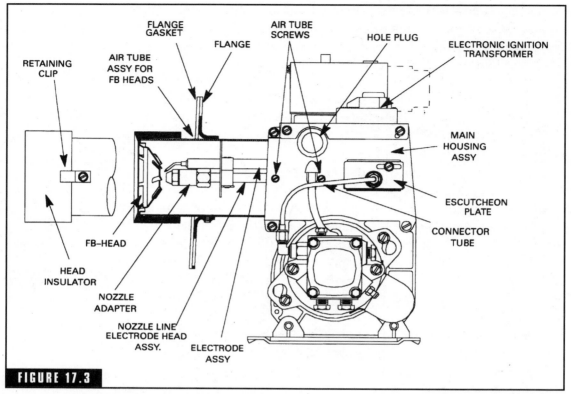

Oil burner components. (*Courtesy of Lennox Industries, Inc.*)

The first item that I like to start with is the gun assembly. This assembly consists of the oil nozzle and electrodes. Both these items are critical to proper operation of the heating system. To remove the gun assembly from the burner assembly, you will need to gain access to this assembly. On some units, there will be an access door that you must remove to gain access. And on others, you will need to move the ignition transformer to gain access. In either case, this is not a difficult procedure. Look at the unit you are working on and make this determination. If you are having a problem locating this assembly, you can find it by following the oil line from the oil pump. This line will be connected to the gun assembly.

1. Remove the access panel, or move the ignition transformer. You will have to remove the oil line from the gun assembly at this time. Next, loosen the retaining nut that holds the gun assembly in place

(Fig. 17.6). On units that you must move the ignition transformer to gain access, the transformer may be connected to the leads of the electrodes in one of two ways.

2. *Direct connection.* This is where the electrodes are connected to the ignition transformer by means of a wire that snaps onto a terminal on the transformer.

3. *Indirect connection.* This is where the electrode has a copper strip connected to the electrode, and this rests on the terminal of the electrode.

If you have the direct type of connection, you will need to disconnect the wire from the terminal of the electrode before the gun assembly can be removed from the burner assembly. Once you have removed the oil line connections, retaining nut, and electrode leads, pull the gun assembly out of the burner assembly.

Once you have the gun assembly removed from the burner assembly, take a few moments to observe the condition of the gun assembly. Look at the nozzle. Is the end of the nozzle burnt or covered with soot? If so, this can be an indication that the air-to-fuel ratio is not set properly. We will look at this more closely when we finish the tune-up and check the operation. Also, look at the electrodes. Are the points of the electrodes dull or pointed? Are there signs of cracks in the ceramic coating? Dull points will not allow as much spark transfer as is needed for the ignition sequence. Pointed electrodes will have a higher concentration

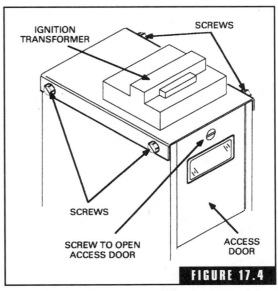

Ignition transformer mounted to box. (*Courtesy of Lennox Industries, Inc.*)

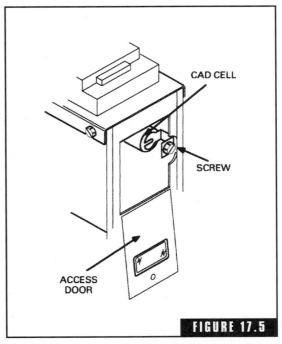

Cad-cell location under transformer. (*Courtesy of Lennox Industries, Inc.*)

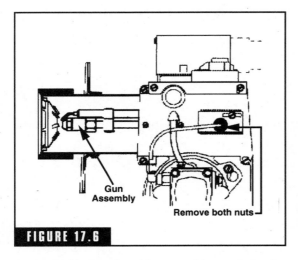

FIGURE 17.6

Gun assembly location. (*Courtesy of Lennox Industries, Inc.*)

of energy than dull ones. You want to make sure that the electrodes have a good point to them. I will show you how to do this.

If the electrode ceramic coating shows signs of cracks (you may have to look very close to see them), the electrodes will need to be replaced. A crack in this coating will allow energy from the transformer to be diverted to areas other than the tips for proper ignition. Electricity will seek any ground that is the path of least resistance. Think of this as a garden hose that has a hole in it. You will not get the pressure at the nozzle that you need because some of the pressure is being lost at the hole. You are also spraying water in an area where you may not want it. It is the same thing here, except you are talking about 14,000 V going to an uncontrolled area. This could be very dangerous. Close examination of the electrodes will tell you if you need to replace them. If you are not sure, play the safe bet and replace them. A pair of electrodes is not very expensive, and you will only be making the forced air heating system operate better by doing this.

You will need to replace the nozzle now. If you have a nozzle wrench, place the wrench over the nozzle and the end of the gun assembly, and turn the end of the wrench that is in contact with the nozzle counterclockwise to loosen and remove it. If you do not have a nozzle wrench, place one adjustable wrench on the flat spot of the gun assembly just behind the nozzle and the other adjustable wrench on the flat spot of the nozzle. Hold the wrench that is connected to the gun assembly, and turn the other wrench that is connected to the nozzle counterclockwise to loosen and remove the nozzle.

Once you have the nozzle removed from the gun assembly, look at the flat spot on the nozzle. There will be a marking telling you the gallon per hour rating of the nozzle and the spray angle. In most cases, you will want to replace the nozzle with a new nozzle with the same gph and angle of spray rating. Use the following calculation to determine if you have the proper size nozzle. Normal U.S. no. 2 fuel oil has a Btu (British thermal unit) rating of 140,000 per U.S. gallon. Therefore, it is easy to calculate the gph rating of the nozzle that you will need. All nozzles are rated for a flow in U.S. gph. If you have a nozzle that is marked 0.75 gph, you will have an output rating of

$$0.75 \times 140,000 = 105,000 \text{ Btus}$$

This calculation will get you very close to the nozzle size that you should use. You will need to check the Btu rating of the heating unit that you are working on by looking at the rating plate that is located on the unit.

If you cannot read the gph rating on the nozzle, you can determine the proper size nozzle to use by determining the Btu rating of the oil forced air unit that you are working on from the rating plate and use this calculation: rating of the heating unit divided by 140,000 (the Btu rating of no. 2 fuel oil). In other words,

$$\frac{105,000 \text{ (unit rating)}}{140,000 \text{ (Btu rating of no.2 fuel oil}} = 0.75 \quad \text{the nozzle size}$$

To maintain the proper efficiency of an oil forced air heating unit, it is important that you know the rating output of the unit so that you

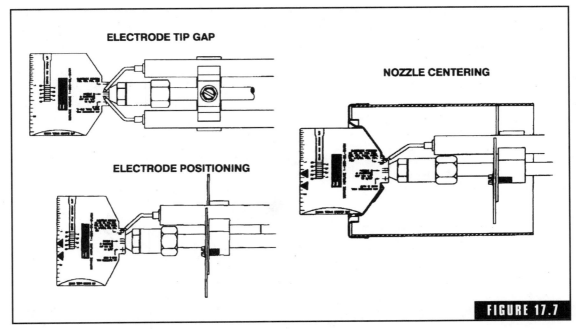

Electrode gauge use. (*Courtesy of Lennox Industries, Inc.*)

do not use a nozzle that is too small (this will cause the unit to work harder than needed) or one that is too large (this will waste oil and cost the homeowner more money). You will not gain anything by putting in a nozzle that will cause a higher Btu output than that for which the unit is designed. The oil pump is only designed to pump so much oil per hour and is sized to the Btu output rating. Therefore, using a larger nozzle will not be of any use. Use the preceding calculation to determine that you are using the proper size nozzle for the Btu rating. Do not assume that the last professional to work on this unit made the proper choice. He or she simply may have replaced the nozzle with one of the same size that was there. At some point a person may have made a wrong choice, and you will simply be following the chain.

It is very important to use the proper angle of spray for the nozzle as well. Once you have determined the size of the nozzle that you need, replace the nozzle with the proper size by placing it in the end of the gun assembly and turning it clockwise to seat the nozzle. Use your nozzle wrench or adjustable wrench to tighten the nozzle into place. Do not overtighten the nozzle because you may strip the threads.

Now that the nozzle is in place, you need to make a decision about the electrodes. Do you need to replace them or simply file the ends to a point? If the ceramic parts of the electrodes are in good shape, you can file the ends to a point. You will need your metal file for this.

You may find that it is easier to remove the electrodes from the gun assembly for this job. There is normally one screw that is attached to a plate in the center of the gun assembly that holds the electrodes in place. You can simply loosen this screw and slide the electrodes out of the holder. Once you have the electrodes removed, file the ends to a fine point. Perform this same procedure on both electrodes. If you are going to replace the electrodes, all that is needed is to make sure that you replace them with a set that is the same length. You also will have to remove the device that connects to the ends of the electrodes that makes the contact either with the ignition transformer or with the wire that snaps onto the terminal. There is a nut on the end of each electrode that you will have to remove to replace the copper contacts, or you will have to unsnap the leads and snap them onto the new electrodes. Once you have either filed the old electrodes or finished preparing the new ones, replace them in the holder. Make sure that the points are facing each other, and only tighten the holding screw enough so that the electrodes can be moved in the holder. The next step is the most critical setting of the burner assembly. You must set the distance between the electrodes, the distance between the electrodes and the center of the nozzle, and the proper distance that the electrodes are up from the center of the nozzle. If these settings are not correct, the unit may not operate properly, or it may not operate at all.

Figure 17.7 shows how to make these settings using an electrode setting tool. If you do not have one of these gauges, it is possible to make the proper settings using a tape measure. Use Fig. 17.7 as a visual aid and see the Quick Tip box for the proper settings.

These settings will put the electrodes in the proper position for the maximum amount of energy to be transferred from the ignition transformer through the electrodes to ignite the fuel. If the gap is too small or too large, this will impede the spark needed for ignition. If the electrodes are too close or too far from the nozzle, there may be a delay in ignition. Thus, you can see that proper positioning of the electrodes relevant to the center of the nozzle is critical to efficient operation of the unit. As a heating professional, you should make every attempt to have an electrode gauge available in your tool box to take the guess work out of setting the electrodes.

Once you have made the adjustments to the electrodes, tighten the screw that is used to hold the electrodes in place, and then recheck the settings in case the electrodes have moved. Loosen the screw and make any adjustments that are necessary, and tighten the screw again. Make sure that when you are tightening the screw you do not over-tighten it, because this can cause the ceramic cover on the electrodes to crack. If this happens, you will need to replace them.

Once you have the nozzle replaced and the electrodes adjusted, you are finished with the gun assembly. Before you put the gun assembly back in place, look down the burner tube where the gun assembly will be located, and examine the end of the burner tube. The end of the tube has a defuser head that can be seen in Fig. 17.4. You want to make sure that this is clear of any debris. I have seen times when this head has a buildup of hard crust that is

QUICK»TIP

1. *Electrode gap.* $\frac{1}{8}$ in between tip of the electrodes.
2. *Electrode centering.* $\frac{1}{8}$ in between tip of the electrodes and the center of the nozzle.
3. *Electrode positioning.* $\frac{1}{4}$ in between the electrodes and the center of the nozzle.

caused by a nozzle that did not use the proper angle of spray that matched the angle of the head. If you see this type of buildup on the end of the head, check to make sure that you are using the proper spay angle nozzle. This information may be on the rating plate, or you can get this information from the manufacturer of the heating system that you are working on. Do not ignore this condition because it can cause serious damage to the heating equipment, and it can cost the homeowner extra money because the heating system is not operating properly. If you observe this condition, you will need to clean this deposit off the end of the burner tube before the gun assembly can be installed. You can use a screw driver or other item to break off the debris and clean it out of the burner tube.

Once this is complete, you will want to check the condition of the induction blower and coupling. This squirrel cage blower is located ahead of the burner tube and is connected to the burner motor. This blower is used to supply oxygen to the burner for proper combustion. If the fins on the cage are covered with dirt and grease, you will need to clean them to allow for the proper amount of airflow. To clean these fins, use a small screwdriver to remove any dirt or debris. Be sure to clean up any debris that falls into the burner tube from the blower fins. Figure 17.5 shows the location of the blower cage.

Once this is complete, check the condition of the coupling. This is the device that connects the burner motor to the oil pump. Make sure that the connections are good and that the coupling is in good shape. If the coupling shows signs of wear, or if the coupling is cracked, it will need to be replaced.

Once you have either replaced the coupling or determined that it is in good shape, you can put the gun assembly back into place. Slide the gun assembly back into the burner assembly, and place the oil delivery tube back into the slot in the burner assembly. Secure this into place with the screw or bolt that was removed. Attach the main oil line fitting to the gun assembly feed tube, and tighten it in place.

QUICK ⟫ TIP

1. Remove the access panel, or remove the ignition transformer.
2. Loosen the set screw that holds the coupling to the burner motor.
3. Loosen the set screw that holds the coupling to the oil pump.
4. Remove the bolts that hold the burner motor to the burner assembly.
5. Slide the burner motor out from the burner assembly.
6. Remove the coupling from the burner motor and oil pump and replace.

If the electrodes were connected to the ignition transformer by means of a direct connection, you may attach the leads to the transformer at this time. If you are planning on checking the voltage of the transformer before you start the unit, do not attach the leads at this time.

If you intend to check the voltage of the ignition transformer for any reason, make sure that you are properly trained and that you have the proper equipment to check this device. The ignition transformer has an output rating of 14,000 V ac. If this device is not checked properly, you can sustain serious injury from that level of voltage. I do not recommend that the homeowner attempt to perform this check. It is better left to the heating professional who has the proper checking equipment.

To check the transformer, you will want to make sure that you are not pumping raw fuel into the fire pot. One way to avoid this is to attach a length of plastic tubing to the bleeder port on the oil pump and open the bleeder nut to divert the oil into a jar or other container to catch the oil. The best time to check the transformer is after you have completed the summer tune-up and just before you are ready to start the unit to check the operation. Thus I will wait until that time to explain the procedure.

If the electrode leads are the copper type where the ignition transformer rests on the leads, you will want to check to make sure that they are in the proper position so that you get the best contact. To check for this, make sure that the leads are high enough to make contact with the terminals on the transformer. Slowly lower the transformer onto the leads, and observe that the contact is made. If you cannot tell, bend the leads up to ensure good contact. You also can place a small amount of grease on either the transformer terminals or the leads and lower the transformer back down and lock it into place. Next, unlock the transformer and move it up out of the way. Look at the leads and the terminals to see that they both show signs that the grease has transferred onto them. If not, adjust the copper strips more until a good contact can be established. Once this is confirmed, wipe the grease from the leads and the terminals, and lock the transformer into place.

If the unit that you are working on is equipped with a cad-cell safety, you will want to remove the cad cell and examine its condition.

Figure 17.8 shows the location of the cad cell. Figure 17.12 shows one type of oil burner control that the cad-cell leads are connected to. This is one of the safety devices used on oil forced air heating systems. (See Figures 17.8 to 17.12.)

The cad cell is used to "see" the light from the burner when the unit is operating. If the cad cell cannot "see" the light from the burner, it will open the circuit and go into a reset mode. The reset button must be pressed to restart the cycle. You can imagine that if the transformer is not able to ignite the fuel, then each time the reset is pressed, more raw fuel will be pumped into the fire pot. The cause of this type of problem will be covered in Chap. 18.

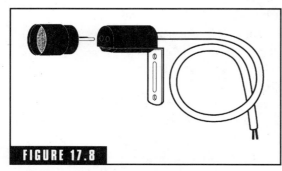

FIGURE 17.8

Cad-cell relay. (*Courtesy of Honeywell, Inc.*)

Remove the access door to gain access to the cad cell. Remove or loosen the screw that holds the cad cell in place, and examine the cell. The cell should not have any dirt or oil on its "eye" that would obstruct the cell from "seeing" the flame. Clean the cell if needed. Take the time to look down the path that the cell must use to check for the flame to make sure that there is nothing in the way of the cell confirming that the flame is on. Once you are sure that the path is clear, replace the cell, and secure into place. Then replace the access door.

Some units are equipped with a stack control instead of a cad-cell relay. In such a case, the stack control will be located in the flue pipe between the chimney and the oil heating unit. Figure 17.11 shows a stack control.

This unit works like a cad cell except that this unit senses the heat generated by the combustion to "see" that the burner is lit. If the unit does not sense this heat, it will shut down the oil pump so that it does not pump raw fuel into the fire pot. If

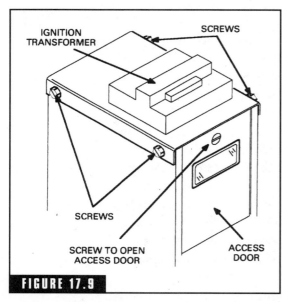

FIGURE 17.9

Ignition transformer. (*Courtesy of Lennox Industries, Inc.*)

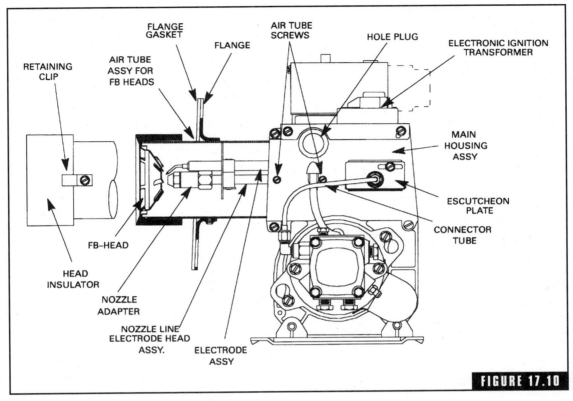

Burner assembly. (*Courtesy of Lennox Industries, Inc.*)

the unit locks out, you will need to press the reset button on the stack control to reset the system.

You have now completed maintenance of the burner assembly, and you can move on to the blower assembly.

Blower Assembly

Now that you have the heating section of the furnace serviced, let's turn our attention to the blower section. As I mentioned at the beginning of this book, there are two basic types of blower units: (1) *direct drive,* where the blower cage is attached directly to the motor shaft, and (2) *belt drive,* where the blower cage is connected to the motor by a belt and pulley system. Both these types of systems perform the same function— to blow warm air into the home during the heating cycle.

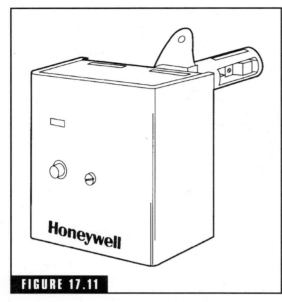

FIGURE 17.11

Stack controls. (*Courtesy of Honeywell, Inc.*)

Locate the blower section of the heating system. If you have determined that you are working on an upflow system, the blower will be located in the compartment *below* the burner unit. If you are working on a downflow system, the blower will be located in the compartment *above* the burner unit. If you are working on a "low boy" system, the blower will be located in the *rear* of the system.

The first thing that must be done before beginning to service the blower unit is to make sure that the power is disconnected to the heating system. Once the power has been disconnected, remove the door(s) to gain access to the blower assembly.

Direct-Drive Unit

If you have a direct-drive unit, look at the end of the motor that is protruding out from the blower housing (Fig. 17.13). Now look at the top of this motor to see if there is a small hole there. If there is, then you will want to add 2 to 3 drops of oil in this hole to lubricate the bearings. In some cases, there also may be a hole on the opposite end of the motor that is inside the blower cage. If there is, you will need to lubricate this area as well. If you can not locate any small oil ports on the motor, then you have a sealed bearing unit, and no lubrication is necessary.

You should now take the time to examine the blower (squirrel cage) as well. If there is a buildup of dirt on the fins of the cage, you need to clean them out. A buildup of dirt on the fins will cause a reduction in the amount of airflow that the blower can provide. You can use a screwdriver to scrape the dirt off the fins, starting from the rear and pulling it toward you. Remove this dirt with either a vacuum cleaner or some other device so that the fins are clean. You also will want to make sure that the blower cabinet is free from dirt and dust. It will not do much good to clean the blower fins and leave dirt in the blower cabinet that will be pulled into the blower the next time the blower is started.

The next item to examine is the filter. In the downflow type of heating system, the most common filter is the *mat-type filter*. In this type of filter, a metal cage holds the filter material. This material typically will come in a roll that is cut to fit the cage. Some heating companies have this material cut to size and prepackaged for the homeowner who wants to replace his or her own filters. The heating professional, though, probably will have a roll of this material on the truck.

Remove the filter cage from the blower cabinet. Slide the bar at each end of the filter cage, and separate the two sides. Remove the old filter material, and measure the length so that you have the proper size. Select the width of filter material that you will need, and cut the new filter to size. You will note that the filter material is a light color on one side and a darker color on the other side. You also should note that one side of the filter material is coated with an oily substance. This is the side of the filter material that *faces up into the cold air duct* (Fig. 17.14). This is the side of the material that will collect the dust and dirt that is pulled into the blower cabinet.

Once you have the filter material cut to size, lay it on the cage and assemble the cage unit. Now install the cage back into the blower cabinet. Replace the door(s) on the blower cabinet. This unit is now serviced and ready for operation.

Belt-Drive Units

If you have a belt-drive unit, this will require slightly more maintenance than a

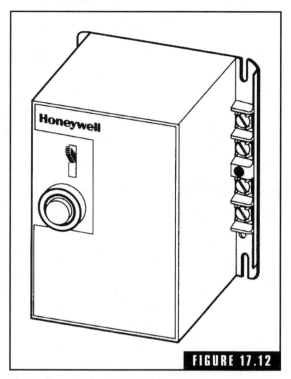

FIGURE 17.12

Cad-cell burner control. (*Courtesy of Honeywell, Inc.*)

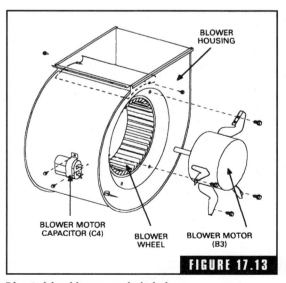

FIGURE 17.13

Direct-drive blower, exploded view.

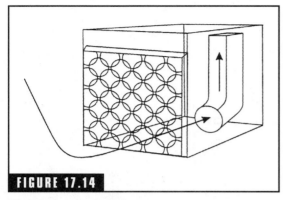

FIGURE 17.14

Filter in return air duct.

direct-drive unit. Most older heating systems use this type of system (Fig. 17.15). On an upflow unit, the blower cabinet will be located under the burner assembly. On a "low boy" type of heating system, the blower cabinet will be located in the rear of the unit.

First, locate the motor unit. This typically will be located on top and to the rear of the blower unit. You will find an oil hole located on each end of the motor. You will need to add 2 or 3 drops of oil into each of these holes. In some cases, where the motor has been replaced by a newer model, there will be no oil ports. In this case, you have a sealed bearing unit, and no lubrication is required. Next, remove the belt from the pulleys. To do this, pull the belt over one of the pulleys, and rotate the pulley clockwise to remove the belt. Care must be taken not to get your fingers caught between the belt and the pulley.

Turn the belt inside out and examine it. If there are signs of splitting or cracking, the belt will need to be replaced. Turn the belt right side out and look for the size of the belt. This will be printed on the outer casing. If you cannot read this information, take the belt to an auto parts or hardware store, and they should be able to measure this for you so that you get the proper size replacement belt. If you are at a service call location where it is not convenient to leave to find an auto parts or hardware store, find a belt in your service truck that is slightly *smaller* than the one that you removed. It can be assumed that the belt has been on the unit for some time and that it has stretched. It is more important to get the proper width of belt to the pulley that is on the unit, since you can adjust for the length if required. You also should look around the heating unit to see if the last person to service the unit left an old belt there. This will give you an idea as to the proper size as well.

While we are on this subject, let me just mention that as a heating professional, you should always leave the parts that you replace with the homeowner. This will serve two important functions:

1. It will show the homeowner that you did in fact replace the parts.

2. The next time that you service the unit, it will allow you to remember what parts have been replaced.

Moreover, you should always write the date of replacement on the carton or package that the part(s) came in and leave them next to the heating unit.

Next, examine the blower unit. In some cases, there will be oil ports on the bearings that the blower shaft goes through. If there are, add oil to these ports. If there are no oil ports, check to see if the blower has grease cups. These would be located on the upper portion of the bearings. If there are grease cups, remove the cup(s) to see if there is enough grease in them. Most of these types of lubricating systems require a special grease that you may be required to purchase from your local heating company. The heating professional should always have at least one tube of this grease on the service truck.

Fill the cups, and replace them on the bearings. Turn them down until you feel a resistance (it becomes harder to turn them). The homeowner should turn these grease cup(s) one-quarter to one-half turn every other month during the heating season to allow for proper lubrication to the bearings. Failure to do this may cause the bearings to go dry. This will cause a metal-to-metal situation that will wear out the blower shaft.

If the heating system has not been serviced for a long period of time, this is one item that you should check as part of the tune-up process. You should ask the homeowner if the blower is noisy when the unit is running. If he or she says "yes," this will give you a good indication that there is excessive wear on the blower shaft.

You can check for this condition in one of two ways:

1. With the belt removed, pull up and down on the blower shaft pulley. If there is excessive play (a good bearing and shaft should not move at all), you have a worn-out bearing.

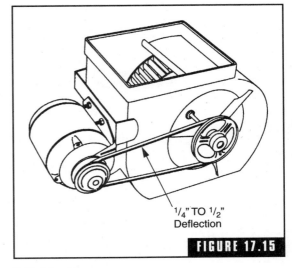

¼" TO ½"
Deflection

FIGURE 17.15

Belt-drive blower.

2. Replace the belt on the motor and blower, and turn the fan switch to "manual" to start the blower. Listen to the sound of the blower. If you hear a squealing noise or a rumble, you have a bad ware situation. You also can look at the way that the pulley turns. If you detect a wobble in the pulley, you have ware.

You need to inform the homeowner of this finding before you proceed any farther. You must get the homeowner's permission to make this expensive repair. In most cases, this will be a time and material job, and depending on how far you are from your shop (you will need to disassemble the unit and go to the shop to repair it), you could be looking at 2 to 3 hours of labor. Explain to the homeowner that the situation will only get worse and that he or she could be looking at this unit failing in the winter, and then he or she would have no heat. Also explain that this is the reason that the homeowner has you come to the home in the summer to tune up the heating system so that the chances of having a failure in the winter are reduced.

Once you have the homeowner's permission to do the repair, you will need to disassemble the unit and remove it from the blower cabinet in the following manner:

1. Check to make sure that the power to the unit is turned off.

2. Remove the locking straps that hold the motor to the blower unit.

3. Remove the motor from the motor bracket, and place it on the bottom of the blower cabinet. Make sure that you do not allow the motor to hang from the wires or conduit that is connected to the motor. Use some kind of brace if needed to support the motor.

4. Remove the bolts that hold the blower unit to the heating unit.

5. Remove the blower unit from the blower cabinet.

Once you get the blower removed from the cabinet, examine the blower fins for signs of dirt buildup. If there is a buildup on the fins, prior to taking the unit to the shop, stop at a local manual car wash and spray down the blower unit. The high-pressure spray works great to remove the old buildup on these types of units. Where the buildup is very heavy, some of these car washes have a setting for degreasers, and this works great.

Once this is complete and the blower unit is at the shop and on the bench, you will need to remove the locking collars that hold the shaft in place. Normally you will find two of these collars located on the opposite end of the blower from the pulley. One should be on the outside of the rear bearing and one on the inside of the rear bearing. On some units, these devices will be incorporated into the blower cage, one at either end. In this case, simply loosen the setscrews, but do not remove them. Loosen and remove the outside collar, and loosen the inside collar. This will then allow the blower shaft to be removed.

If the bearings have been neglected for a long period of time, the shaft may be "frozen" to the bearings and will not slide out of the unit. In this case, lubricate the entire shaft with motor oil. You will then need to use something to drive the shaft out of the bearings. Never strike the end of the shaft with a hammer or other driving device. This may cause the end of the shaft to flare. The best approach is to use a piece of shaft material that is either the same size or slightly smaller to drive the shaft out of the bearings. Once you have the shaft free from the rear bearing, you can attempt to pull it out the remainder of the way by twisting the shaft or by turning the pulley back and forth while pulling at the same time. If this works, pull the shaft free from the blower unit. If the shaft still will not come free, you will need to continue with the driving method until the shaft is free from the blower unit.

Once the shaft is free, you will need to remove the pulley from the shaft because you will need to use this pulley on the new shaft. Loosen the setscrew that holds the pulley on the shaft, and remove it from the shaft. If the pulley will not come off, place motor oil between the end of the shaft and the pulley, and use a wheel puller to remove the pulley. Place the jaws of the puller around the pulley, and screw the jacking bolt down to make contact with the end of the shaft. Tighten the jacking bolt with a wrench until the pulley comes free. Care must be taken if the pulley is aluminum that you do not bend it. If it does bend, you will need to replace the pulley as well.

Next, remove the bearings from the blower unit, and find replacements that are the same number. It does not matter if the replacement bearings have grease fittings or oil fittings; they will both work equally well. I prefer grease fittings over oil fittings because it is much easier to get the homeowner to turn the grease fittings down than it is to get him or her to place oil in the oil fittings.

Once you have the new bearings installed, you will need to make a new shaft. To do this, you will need to take three measurements of the shaft:

1. Measure the length of the shaft.

2. Measure the diameter of the shaft.

3. Measure the length of the flat spot on the shaft that is used to tighten the pulley.

Once you have all three of these measurements, cut the shaft to length, and grind the flat spot. Lightly oil the shaft and slide it into the first bearing, through the blower cage, and insert the first locking collar. Next, slide the end of the shaft through the rear bearing, and attach the other locking collar. Tighten the locking collars only finger-tight because you will have to make the final adjustments once the blower is located back in the blower cabinet. In the case where the setscrews are part of the blower cage, use this same procedure and only tighten them finger tight as well. Slide the pulley onto the shaft, making sure that the setscrew is lined up with the flat spot on the shaft. Tighten the setscrew finger tight as well.

Once the blower is reinstalled in the blower cabinet, replace the motor on the motor bracket, and secure it in place. You will need to measure the distance from the outside of the blower housing to the inside of the motor pulley. This measurement will need to be the same as the distance from the blower housing to the inside of the blower pulley. This distance is critical so that the belt will travel in a straight line from the blower motor to the blower itself. If the belt does not travel in a straight line, the belt can either jump off the pulley or cause excessive wear in the new bearings that you just installed.

Once you have the alignment correct, tighten all setscrews on the blower shaft and the pulley. Also double-check to make sure that the mounting bolts for the blower unit and motor are tight. Now check the belt tension, and make sure that you have the proper $1/4$- to $1/2$-in deflection. Make the necessary adjustments to the motor adjustment jacking screw to achieve this deflection.

Turn on the power to the heating unit, and manually start the blower only by means of the blower switch or by adjusting the fan control to start the blower. Watch how the belt turns on the blower pulley. If there does not appear to be a straight line of travel between the motor pulley and the blower pulley, turn off the power and adjust as necessary; then recheck it. When the alignment is correct, this repair is complete.

You will now want to examine and/or replace the air filter. The air filter will be located either inside the blower compartment or inside the cold air duct leading to the blower compartment. Filters come in several types depending on the type of heating system that you are working on. They will be either a box type, where the filter material is enclosed in a cardboard box and has a mesh cover on the side that faces the blower, or the filter material will be enclosed in a metal frame. Filters come in several different sizes as well. Locate the filter in the heating unit, and remove it. If you have a box-type filter, the size of the filter will be imprinted on the outside of the filter. Replace this filter with the same size filter. Make sure that you look for the arrow on the side of the filter that will show you which way the filter is to be installed. Typically, the way to remember is that the side with the metal mesh that covers the filter material will face the blower compartment. This is so that the filter material is not sucked into the blower during the heating cycle.

If you have the type of filter that is mounted in a removable frame, you will need to measure the amount of filter material you will need. This can be done by measuring the existing material that is mounted to the frame. Write this information down for future use. Cut the proper size filter from the roll, and install it on the frame, making sure to put the coated side out. Place the other side of the frame on the filter, and lock it into place. Reinstall the filter frame into the blower cabinet.

The summer tune-up of the oil forced air heating system is now complete. All that remains is to replace any doors and access panels that you removed during this operation and turn the power back on by resetting all breakers or replacing the fuse(s). The final step is to check the operation of the heating unit by running it through a complete heating cycle and checking the safeties.

Before you turn on the power to the oil forced air heating unit, you will need to disconnect the power to the blower unit. The reason for this is so that you can check for the proper operation of the high-limit safety. Figure 17.16 shows one type of fan control. This is the high limit that is set on the fan control and will shut off the power to the burner motor and blower if the blower does not come on during the heating cycle.

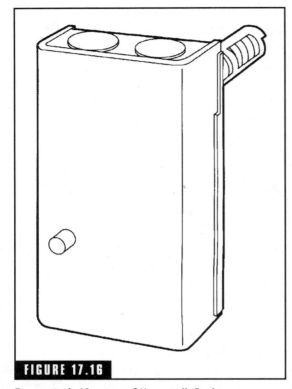

FIGURE 17.16

Fan control. (*Courtesy of Honeywell, Inc.*)

If you are working on a direct-drive unit, open the access panel to the wiring cabinet (make sure that the power is turned off first), and locate the black (hot) wires. Look at the wires, and you should find a wire marked for the blower. Remove the wire nut that holds all the wires together, and remove the wire for the blower. Replace the wire nut on the remaining wires, and make sure that the black wire for the blower is not touching any metal objects. You also must make sure that you do not touch this wire while the power is turned on. If you have a belt-drive blower, you will not have to disconnect any wires. Simply remove the belt to simulate this same condition.

Before you turn the power back on to the unit, you also need to check the thermostat to make sure that it is operating properly. Remove the front cover from the thermostat, and check for any lint or dirt. Remove any, if present. Next, turn the thermostat up *slowly* until either you see the mercury bulb drop to the right to indicate that the thermostat is calling for heat or you hear the bimetallic points come together indicating the same thing. Now look at the pointer on the top of the thermostat that shows what the reading is when the heating cycle starts. Now compare this to the reading on the lower part of the thermostat (the thermometer reading that shows the actual temperature in the home). These readings should be the same. If they are not, you need to adjust the thermostat so that the two readings are the same.

If you have a thermostat wrench, place it on the nut behind the coil that controls either the mercury bulb or the points. Determine from the readings if you needed to turn the coil to the left (thermostat set point higher than the actual temperature) or to the right (thermostat set point lower than the actual temperature). Make very slight movements with the wrench. After you have made the adjustment, turn the thermostat all the way down (to the left) and then up (to the right) until the thermostat is calling for the heating cycle to begin, and compare the readings. Make the necessary adjustments again until the readings are correct. Once this is done, leave the thermostat turned up (calling for heat), and return to the heating unit.

Turn the breaker on or replace the fuse so that there is power to the heating unit. The burner motor should begin to operate, and the ignition transformer should ignite the fuel. Once the heat has reached the set point on the high limit, the heating unit should shut down. If it does not, you will need to replace the fan control, since the safety feature on this device is not working properly. This could cause a major overheating situation that is unsafe.

To replace the fan control, shut the power off to the heating unit, and remove the wires to the fan control. Write down the location of these wires if needed for reference when you install the new one.

Fan controls come in two basic types

1. The first type has a round metal tube with a metal coil inside that contracts when heated causing the dial on the front of the fan control to turn until the set point is reached to send power to the blower motor, causing it to run.

2. The second type has a bimetallic strip that comes together when heated to send power to the blower motor, causing it to run. The front of this type of fan control has a sliding lever to adjust the set points.

Both these fan controls work the same way, and they come in different lengths. It is important that you do not put in a new fan control that is a different length from the one that you are replacing. These controls come in different lengths depending on the size of the heating unit in which they are used. If you use one that is too long, it may make contact with a metal surface inside the heat exchanger and give a false reading and therefore not operate properly. One that is shorter will not operate properly because it will not be able to reach far enough into the heating stream to be effective.

Remove the screws that hold the fan control to the heating unit, and remove the unit. Once you have determined the proper fan control to use, install the new fan control and reattach the wires. Adjust the off set point to between 70 and 90°F and the on set point to between 140 and 150°F.

Turn the power to the heating unit back on, and check the high limit again. The heating unit should shut down when the fan control reaches this number. If this fails again, recheck your steps, and try again. You may not have connected the wires properly, or you may have used the wrong fan control for this type of unit.

Once this check is satisfactory, reconnect the black wire for the blower (make sure that the power is turned off), or replace the belt. Now start the heating cycle again. This time you are checking to make sure that the heating system operates properly for a complete heating cycle.

If you replaced the fan control, you are looking for one more item. This is when the blower is running, in a normal heating cycle, that it remains on during the complete cycle. If the blower starts, runs for a period of time, and then shuts off while the burners are still on, you will need to adjust the fan-on-temperature higher to allow more heat to build up prior to the blower starting. If the blower shuts off at the end of the heating cycle and then comes back on again, you will need to adjust the off temperature lower to allow for the blower to run longer so that there is no more heat buildup in the heat exchanger prior to the blower shutting off.

Once you have checked the safeties and are sure that everything is working properly, you will want to check the temperature rise of the unit. Each heating unit has a temperature range that is printed on the rating plate. To check this range, you will need to place one thermometer in the warm air (supply side) ductwork and one in the cold air (return side) ductwork (Fig. 17.17). You must make sure that the thermometer that is placed in the warm air side cannot "see" the heat exchanger, thus picking up the radiant heat.

Set the thermostat to the highest setting, and put the thermometers in place (you will have to make small holes in the ductwork if they are not already there). Once the thermometers have reached their highest steady temperature, subtract the two readings to get the rise. If this reading is within the range on the rating plate, this step is complete. If the reading is too high, you will need to speed up the blower. If it is to low, you will need to decrease the blower speed. On direct-drive units, you will need to select the proper color wire to make the adjustment. On belt-drive units, you will need to loosen the setscrew in the end of the pulley. Turn the pulley clockwise to decrease the speed and counterclockwise to increase the speed. Tighten the setscrew.

After you have made the adjustment, check the readings again, and adjust the speed as necessary to get the proper reading. There is one more critical check that we need to talk about before the summer tune-up section is complete, and this is the heat exchanger check.

The heat exchanger needs to be checked to make sure that there are no holes in it. The heat exchanger's job is to convert the heat generated from the burners into heat for the home while diverting harmful gases (carbon monoxide) created from the combustion process out the chimney and away from the home. If the heat exchanger becomes old and worn, holes can develop that will allow this gas to escape into the home. Carbon monoxide is a colorless, odorless, tasteless gas, so it is almost impossible for the homeowner to detect until it is too late. No summer tune-up or, for that matter, emergency heating call should be considered complete until you are absolutely sure that there is no problem with the heat exchanger.

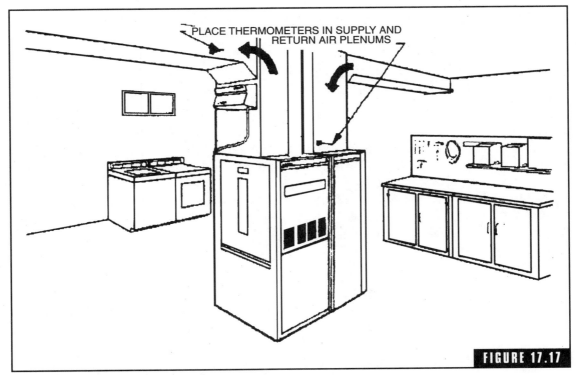

Temperature rise thermometer placement. (*Courtesy of Lennox Industries, Inc.*)

There are two ways to check for the proper operation of the heat exchanger:

1. Observe the burners on the heating system when the blower starts. If you notice that the flame reacts in a different way or that it becomes "lazy" and blows around, you have a problem with the heat exchanger.

2. Make an inspection hole in the hot air plenum that is connected to the heating section of the unit, and using an inspection mirror, look inside during the heating cycle when the burners are on. If you can see light coming from anywhere in the heat exchanger, then there is a danger.

In either case, you *must* inform the homeowner of your findings and red tag the unit. Turn off the oil supply, and inform the company that you work for.

In either case, the homeowner has two choices to make:

1. Replace the heat exchanger.

2. Replace the heating system.

This is a choice that the homeowner *must* make. You cannot allow a heating system with a faulty heat exchanger to continue to operate because it can result in death if all the conditions are right.

It is always a good idea to mention to the homeowner of an oil heating system that it is advisable to purchase and install a carbon monoxide detector in the home as a safety precaution. The cost is inexpensive, and it may help to save a life.

If everything checked out as described and you did not discover any other problems, the summer checkup of your heating unit is complete. If you do not want to wait for the thermostat to turn off the burner unit, you may turn it down at this time and allow the blower unit to complete the cycle. If the heating unit did not function as described above (e.g., burner did not ignite, blower did not come on, etc.), go back over the steps again to see if you missed anything. As I mentioned earlier in this section, you may uncover a more serious problem with your heating system during this process, but it is better to find out now than when you have to use your heating system for the first time and it does not work.

If you have completed all the steps described in this chapter and the heating system is operating properly, you are finished. You should have acquired the confidence at this point to move on to the next chapter on troubleshooting an oil forced air heating system in the event of a failure during the heating season.

Troubleshooting an Oil Forced Air Heating System

Troubleshooting oil forced air heating systems is, in my opinion, the most challenging task described in this book. Oil forced air heating systems have more parts that can fail and have settings that are most critical in terms of accuracy. One of the items that makes this such a difficult and challenging heating system is that it is the only system that burns a fuel that is stored outside the home in liquid form. As I mentioned in Chap. 15, if the tank is not properly maintained, it can cause heating system failure in the winter. When this happens, you may be called out to thaw the oil line. This is not a fun project when the temperature is $-10°F$ and the wind is blowing. However, this is just one of the many causes of failure in this type of heating system.

As with all the forced air heating systems in this book, a systematic approach to the problem will save you a lot of time in making your determination of a solution. We will begin with the most common of the heating problems in the winter and the most common call that any heating professional will face—no heat.

No Heat

When you first arrive at the client's home, you must talk to him or her because the client is your best source of information on where to start. Since every problem has a specific symptom, this will aid you in the best place to start. Listening and observing will be your best tools.

The first item to check is the thermostat. Take a look to see if the thermostat has a heat–cool–off setting on it (Fig. 18.1). If it does, make sure that the setting is for heat. If it was not set for heat, make the change, and listen to see if the heating system starts. Also make sure that the setting for the room temperature is set higher than the temperature in the room. If this setting is too low, raise the setting to be higher than the room temperature. Does the heating system start? If so, let it run through a complete cycle to make sure that all is working properly. If this was not the problem, you will need to turn your attention to the heating system. Use Fig. 18.2 as a reference for the parts of the burner assembly.

Next, you want to make sure that there is power to the unit. This will be controlled by either a fuse or a breaker. Locate the fuse or breaker panel. Open the front panel, and check to see if there is a marking on the panel for the fuse or breaker that controls the heating system. If there is, check the condition of the fuse or breaker. Is the fuse in good condition, or does it show signs of a blown fuse? If you cannot tell, locate the fan control on the heating system. This will be a silver box that is mounted either on the front or sometimes on the side of the unit. This control also can be a rectangle black control with a slide control.

Once you locate the fan control (Fig. 18.3), set the control so that the fan will run on manual. This is done by either

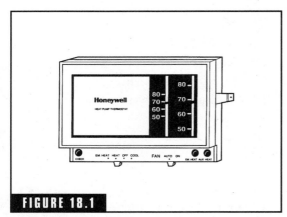

FIGURE 18.1

Thermostat with on–off–fan settings. (*Courtesy of Honeywell, Inc.*)

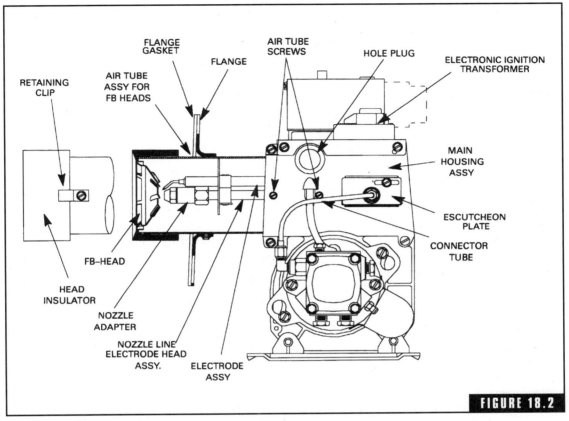

Side view of the oil burner assembly. (*Courtesy of Lennox Industries, Inc.*)

pulling the knob out to the "manual" position or moving both levers to the lowest setting. Before you move the levers, check the location of the levers prior to moving them so that you can return them to the proper position after the check is done. You also can check the thermostat to see if there is a manual fan setting on it. If there is, set this control to "manual" to simulate the same check.

If, for some reason, the blower does not start, there may be another switch that has a fuse attached to it that has not been checked. This switch may be located on the unit, or it can be mounted on a floor joist in the basement. Once you have located this switch, replace the fuse, and check the operation of the blower. It should be running at this point.

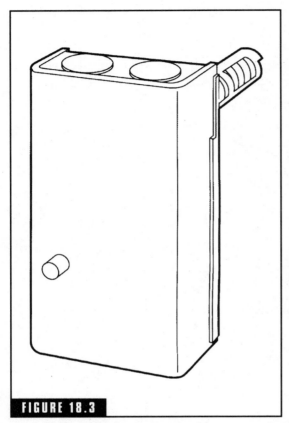

Fan control with manual fan switch. (*Courtesy of Honeywell, Inc.*)

One thing to mention about fuses, when you have to replace a fuse, you should replace it with one of the same ampere rating. You also should choose a fuse that is a slow blow or delayed fuse. These types of fuses allow a delay in blowing to allow for the high-amperage draw when a motor starts.

Now that you have established that you have power to the unit, you need to look at the other items that can cause no heat. There are only three things that will cause a forced air heating system not to operate. You have already checked the first cause—power. The other two are fuel and ignition. Without these three elements present, the unit will not operate. It is important to understand the operation of an oil forced air heating system as described in this section so as to understand how to properly troubleshoot the heating system.

1. If the forced air heating system does not have any power, nothing will operate.

2. If the forced air heating system does not have fuel, it will not operate.

3. If the forced air heating system does not have a source of ignition, it will not operate.

To properly troubleshoot an oil forced air heating system, you should follow this sequence of events to lead you to the solution of the problem. Since you have already covered how to check for power to the unit, and assuming that there *is* power to the unit, let's look at the next possible cause—fuel.

The first thing to do to check for the presence of fuel is to check the tank. Go to the oil storage tank and look at the gauge (Fig. 18.4). Is there a reading on the gauge? If you cannot see a reading, bang on the side of the tank to see if you can notice a difference in the sound as you move down the side of the tank. If the sound is loud, you need to ask the homeowner when fuel was last delivered. If the answer is that he or she has not had a delivery in some time, you may have found the problem. The problem also can be that the tank is so low on fuel that the pump is only picking up the water and sediment that is at the bottom of the tank. If this is the case, after the homeowner has the fuel delivered, you will want to change the oil filter and nozzle because they will have these deposits in them as well. Have the homeowner call you for service after the delivery is made. There is no more that you can do at this time.

However, if you check the tank and there is oil, then move on to the next step. Once you have made the determination that there is fuel in the tank, you need to determine if the fuel is getting to the burner.

Take a look at Fig. 18.5, and find the location of the bleeder on the oil pump. You are going to use this bleeder to make the determination that one of the following is happening:

1. You have a good delivery of fuel, and you can eliminate this as a possible problem.

2. The flow of oil is low, meaning a possible plugged line, tank filter, or oil pump.

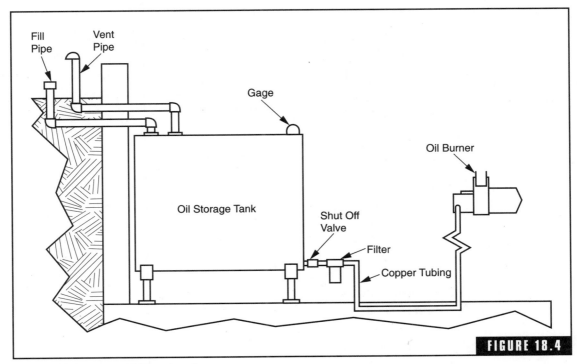

Oil tank with gauge and filter location.

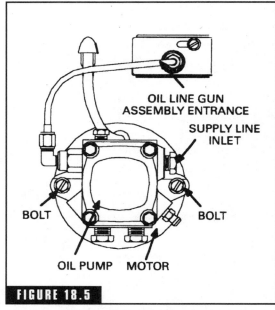

FIGURE 18.5

Oil pump. (*Courtesy of Lennox Industries, Inc.*)

OIL LINE GUN
ASSEMBLY ENTRANCE

SUPPLY LINE
INLET

BOLT

BOLT

OIL PUMP   MOTOR

3. The oil has a "milky" look to it.

4. There is a loud hissing sound with very little fuel delivery.

The first step is to find a device to catch the oil that will be coming out from the bleeder once you start the unit. I like to use a clear jar so that I can see the condition of the fuel. It is also a good idea to attach a piece of plastic tubing to the bleeder to better direct the fuel flow into the jar.

Once you are set with something to collect the fuel and a length of plastic tubing, place the tubing over the bleeder and the other end in the jar. Find the open-end wrench that you will need to open the bleeder. Once you are set, turn the power on to the unit. If the burner motor does not start, you will have to press the reset button on the stack control or cad-cell relay. Once the burner motor is running, crack the bleeder valve open enough to get fuel flowing from the tube. This is where you have to make the determination from the preceding list. You will cover them one at a time, from start to finish, so that you can make the needed repairs.

Fuel Looks Clear and Has a Good Flow

If the fuel is clear and flowing freely, then the problem is not with the fuel. Close the bleeder valve while the pump is still running to keep the prime and so that air does not enter the system. The problem in this situation will have something to do with ignition. This problem will be addressed once all the possible fuel situations are covered. If the fuel is not the problem, skip to the section on ignition later in this chapter.

If the fuel is not flowing at a steady rate but looks clean and clear, you have a problem with a plugged oil line, tank filter, burner motor, or oil pump. When you find yourself in one of these situations where there can be more than one cause, choose to check the simple items first, before moving on to the more complex. In some cases, the simple answer is the correct one. Let's check out the oil line and tank filter first.

You have determined that you have fuel to the pump, so if the problem is in the line, it will have to be between the tank and the pump. Since the oil filter is, in most cases, located close to the tank, let's remove the oil filter for inspection and check the flow of oil from the tank. Refer to Fig. 18.4. Close the valve at the tank. Loosen the bolt that is located on the top of the oil filter until the filter is free to be removed. Look at the condition of the filter. Is there dirt and sludge in the filter. If so, you will want to replace the filter after you check the flow of oil from the tank. If the filter shows signs of dirt and sludge, this would be a possible cause of the problem. Clean the bowl that holds the filter before replacing the filter. It is still a good idea to check the flow of oil from the tank while you have the filter removed.

With the filter removed, place a pan under the filter opening, and turn the valve back on to the tank. Is there a good flow of oil from the tank? If there is, this is not the problem. If the flow from the tank is slow or none at all, you will need to inform the homeowner that he or she will need to contact the company that they receive their fuel oil from to have them check the tank. The homeowner may have to have the tank pumped out to remove the sludge that is causing the problem.

Even if the filter was in good shape, and you have a good flow from the tank, you may have a problem in the line from the tank to the heating unit. You need to consider the outside temperature at this point. If it is extremely cold, you may have ice forming in the line assuming that the line is exposed to the elements and is not insulated. Use your torch to thaw the line before you put the filter back on and attempt to get a good flow of oil from the tank.

Once you have made the checks of the filter and tank, and thawed the line if needed, place the filter (or new filter) back into the holder. Place the gasket on the rim of the bowl if needed to make sure that you have a good seal. Once you have the bowl in place, and the bolt tightened, open the valve on the tank to allow oil to flow. Open the bleeder on the top of the filter housing (the small nut on the top) to allow the air to be removed from the line. You may need to "wiggle" the nut until you have a good flow of oil coming from the nut. Once you have a good flow, close the nut and tighten. Go back to the heating unit and place the plastic tube in the jar one more time. Turn the power to the unit on, and start the pump. Open the bleeder valve and check the condition of the oil. You should now have a good flow of oil coming from the tube. If you still do not have a good flow, you will need to check the line from the tank again. You may have to take more time thawing the line. If the temperature is not extremely cold, and you had a good flow at the tank, and the filter is in good shape, then you may have to replace the line from the filter to the heating unit. If this is the case, you should check with your local heating company as this is best done by the installers, not the service professionals and will not be covered in this book.

Once you have established a good flow of oil from the pump, with the pump still running, close the bleeder valve to divert the oil back to the gun assembly. If this was the cause of the problem, the burner will light at this point. Run the unit through a complete cycle to make sure that the problem has been taken care of.

Oil Has a Milky Look to It

A milky look to the oil is a sign that air has entered the system. As the oil delivery part of the oil forced air heating system is a "closed" system, any air in the system will cause the flow of oil to be disrupted, and the unit will not operate. Another good sign of this is to listen to the sound of the oil pump when it is running. If there is oil in the system, the pump will "whine" while it is running. This is due to the fact that it is not picking up the suction from the tank that a good oil flow will have. This is a good sound to listen to at this point so that you can recognize it if you hear it again so that it will cut down on the amount of diagnostic time needed to discover that there is air in the system.

Air will enter the system any time that there is a break caused in the oil delivery line. This can be caused by:

1. The oil tank running out of oil, and the system not being purged after the delivery.

2. The oil filter was changed, and the system was not bled.

3. The seal on the filter housing is bad and allowing air into the system.

4. One or more of the fittings on the oil line is loose or cracked.

These are the most common causes of air in the system. In each of these cases, the cause must be determined and corrected before the system will operate. It is a good idea to talk to the homeowner before moving on to see if the tank did run out. If it did, then all that is needed is to bleed the system at the oil filter, then to bleed the system from the oil pump bleeder. If the tank has run dry, it would also be a good idea to check the condition of the oil filter at this point as we are going to have to bleed the system anyway. The chance that debris has entered the filter after a complete fill is pretty good. The cost of an oil filter is small compared to having to come back and replace it later if it did become plugged as a result of the fill up and you did not check.

Replace the filter as we have described. Be sure to replace the gasket at the same time. Turn the valve on at the tank and bleed the filter with the small bleeder screw on the top of the filter housing. Continue to bleed the system until you have a good flow from the bleeder. Close the bleeder and return to the heating unit. Turn the power back on to the unit, if needed and start the pump. Open the bleeder on the pump until all of the air is out of the system and you have a good flow of oil. It may be necessary to restart the unit a couple of times to get all of the air out of the system. Once this is done, close the bleeder and the burner will light. Run the system through a complete cycle to make sure that the problem has been solved. If the burner fails to light, there is a chance that some of the debris has made its way up to the filter on the nozzle and is restricting the flow of oil at that point. You should open the inspection door on the front of the unit, and look inside with a flashlight or trouble light to see if you can see the oil coming from the nozzle. If you cannot, you should change the nozzle. You will have to remove the gun assembly from the burner assembly to change the nozzle. To do this, turn the power off to the unit. Remove the cover to the rear of the burner assembly, or on some units, remove the screws that hold the ignition transformer in place, and move the transformer. Remove the oil supply line to the gun assembly. Loosen the retaining nut that holds the gun assembly to the burner assembly. Figure 18.6 shows the location of these parts.

Remove the transformer leads from the ignition transformer if so equipped. Slide the gun assembly out from the burner assembly. Use a nozzle wrench if you have one to remove the nozzle. If you do not have a nozzle wrench, use two adjustable wrenches to remove the nozzle.

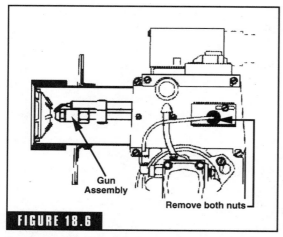

FIGURE 18.6

Gun assembly. (*Courtesy of Lennox Industries, Inc.*)

Before you replace the nozzle, it is a good idea to check to make sure that you are using the proper nozzle for the Btu rating of the unit. To determine the size of the nozzle, take the rated Btus of the unit and divide that number by 140,000 (the amount of Btus in a gallon of #2 heating oil). As an example, a forced air oil heating unit, that has a rated Btu output of 105,000 Btus would use a .75 nozzle. Also make sure that you check the angle of the spray. This is the second number that is marked on the nozzle. This is the angle of the spray that the burner heat is designed to use. So a nozzle that is marked .75 80 would mean that it will deliver .75 gallons of fuel per hour at a spray angle of 80 degrees.

While you have the gun assembly out, it would be a good idea to check the condition of the electrodes as well. Any time that you have the gun assembly out, you should check this as well. As you will learn in the section on ignition problems, if the electrodes are not set properly, there is a good chance that the unit will not operate properly, or at all.

Use an AF11 electrode gauge to set the proper gap and distance (Fig. 18.7). If you do not have a gauge, you can set these gaps as follows:

1. The distance between the tips of the electrodes should be ⅛ in.

2. The distance from the center of the nozzle to the electrodes should be ⅛ in.

3. The electrodes should be ¼ in above the center of the nozzle.

These settings will work on most units and are a good starting point if you do not have the AF11 gauge.

Once you have the electrodes set, and the nozzle installed, reinstall the gun assembly into the burner assembly. Tighten all fittings and locking nuts. Replace the cover, or put the transformer in place. If the electrodes were directly connected to the transformer, connect them before closing the transformer. Turn the power back on to the unit, and start the cycle. If everything else is working fine, the burner should light. Run the unit for a complete cycle to make sure that everything is working properly.

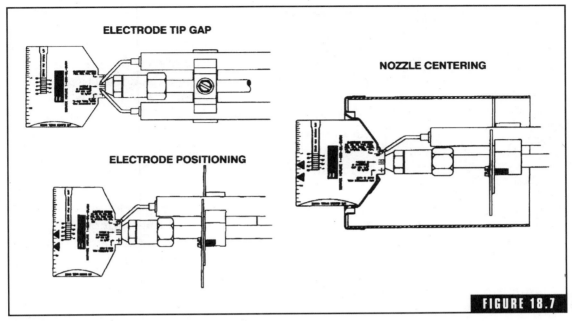

Electrodes and gap. (*Courtesy of Lennox Industries, Inc.*)

If you are still getting milky oil, you may have a cracked fitting or flare. Turn the power off to the unit. Turn the oil valve off at the tank. Loosen and remove the supply side line at the pump. Slide the nut down and look at the flare on the end of the line. If you see any signs of cracking, you will need to remove the bad section and flare the end of the line. Also check the condition of the nut. In some cases, this is the cause. If the threads have become worn, or the nut was "cross thread," then this is also a possible cause of air entering into the system. If either of these are the case, cut the line with your tubing cutters, and remove the nut. Place a new nut on the line. Use your flaring tool to make a new flare in the end of the line. Reattach the line and tighten the nut.

Turn the tank valve back on, and the power to the unit. Start the pump and open the bleeder on the pump and bleed the air from the system. Close the bleeder with the pump still running to allow oil to flow to the gun assembly and begin the cycle. Allow the unit to complete the cycle to make sure that the problem is solved. If fuel was not the problem, then we need to look at the ignition part of the system. This will consist of two potential problem areas. The ignition transformer, and the electrodes.

Ignition Transformer

If the fuel was not the problem, there are two more areas that must be checked. The ignition transformer (the device that supplies the power to the electrodes) and the electrodes (the device that converts the electrical power to the spark needed for ignition). First, let's look at the transformer. Figure 18.8 shows one style of transformer. This unit may be mounted on the top of the burner assembly, or it can be mounted on the side of the burner assembly. In either case, this is a very high voltage transformer. It is sometimes referred to as a step up transformer as it takes the 110 V ac power and will increase the power output to 14,000 V ac. You must use extreme caution when working with this unit. For this reason, I do not recommend that a homeowner even attempt to check this unit for power. This should be left to the heating professional who has the proper equipment to check this device.

To check the transformer, first make sure that you have turned the power off to the unit. Next, loosen the screws that hold the transformer in place. On the top mount unit, lift the transformer to tilt it out of the way. On this type of transformer, the unit is on a hinge and will tilt out of the way to expose the contacts. On the other type of transformers, you may need to remove an access door to gain access to the contacts. You may also have to disconnect the electrode leads from the transformer.

Turn the power back on to the unit. You will need to use a transformer tester to check the amount of voltage. Start the unit by either pressing the reset, or turning the thermostat up. If you must turn the thermostat up, turn the power back off to the unit, turn the thermostat up, and then return to the heating unit before turning the power back on. The reason for this, of course, is that you have determined that there is not a problem with the fuel sup-

FIGURE 18.8

Ignition transformer. (*Courtesy of Lennox Industries, Inc.*)

ply. Every time that you start the unit, you will be pumping raw fuel into the fire pot. This fuel will have to be burned off once you have the problem fixed.

Turn the power back on to the unit and place the probe of the tester on one of the contacts. Turn the dial clockwise until you have a reading of the amount of voltage output of the transformer. If the reading is good, the problem is not with the transformer. If the reading is bad, you will need to replace the transformer. Do not attempt to check the transformer output by shorting a screwdriver across the contacts. You can sustain serious injury, and possibly damage the transformer.

If the transformer checks good, you will need to turn your attention to the electrodes. This will be discussed in the next section.

To replace the transformer, turn off the power to the unit. Locate the box on the burner assembly that houses the wire connections. This can be done by tracing the wires from the transformer. Disconnect the two wires from the transformer. Pull these wires free from the box. Remove the screws that hold the transformer to the burner assembly. Remove the old transformer. If this unit is hinged, you will have to remove the hinge from the old transformer to adapt it to the new transformer.

On the older units, with the stationary transformer, place the new transformer in place and secure with the screws. Run the wires into the wire box and make the connections. Reattach the leads from the electrodes.

On the units that have the transformer hinged to the top of the burner assembly, replace the hinge on the new transformer. Do not tighten the screws all of the way as you will need to adjust the hinge for the proper fit. Attach the screws from the hinge to the burner assembly. Route the wires to the wire box and make the connections.

Close the transformer and adjust the transformer until you have a good fit on the top of the burner assembly. Make sure that you have the proper alignment with the electrode contacts. You may have to place a small amount of grease on the leads so that you can see if you have good contact. To do this, place the grease on the electrode leads, and close the transformer. Press down on the transformer so that you have good contact with the leads. Open the transformer and see if you have grease on the transformer contacts indicating good contact with the electrode leads. If you do, clean the electrode leads and transformer contacts, and tighten the hinge screws on the transformer. If you do not show a good contact, adjust the hinge and repeat this procedure until you show good contact. Close the transformer and secure in place.

Open the inspection door, and turn the power to the unit back on. You may have to press the reset button to get the unit started. Once the unit is started, the burner should light. If the burner does not light, recheck your steps including checking the output of the transformer in case you have not wired it properly. Also, check the contact with the electrodes. It does not do any good to replace the transformer if it is not making contact with the electrode lead. This would be equal to replacing the spark plugs in your car and not connecting the leads from the spark plug wires to the spark plugs and then wonder why it will not start.

Once you have the burner running properly, turn the power off to the unit to allow the excess oil to burn away. You will want to turn the thermostat down and then turn the power back on so that the blower can keep the unit from overheating while the oil burns off. Once the excess oil has burned off, run the unit for a complete cycle to make sure that the repair has been done successfully.

Electrodes

If the transformer checks out and shows that it is putting out the proper voltage, the problem may then be traced to the electrodes. To check the electrodes, we will have to remove the gun assembly. Figure 18.6 shows the gun assembly.

To remove the gun assembly, turn the power to the unit off. Remove or tilt the transformer out of the way, or open the access door that leads to the assembly. Loosen the oil supply line to the gun assembly that comes from the oil pump and remove from the gun assembly. Loosen the locking nut that holds the gun assembly in place. If the electrodes are connected to the transformer by wires, remove these wires from the transformer. Slide the gun assembly out of the burner.

With the gun assembly out of the unit, examine the condition of the electrodes. You are looking for the following conditions:

1. What is the condition of the tips? Do they have a fine point or rounded point? Rounded tips are a sign of worn electrodes. The electrode should have a fine point to better direct the electricity from the transformer for better ignition.

2. What is the condition of the electrode ceramic coating? You are looking for any signs of fine cracks in the ceramic. Any cracks in the ceramic coating will cause a loss of energy to the tips of the electrodes.

3. What is the condition of the contacts or wires connected to the end of the electrodes? As this is the main contact that the electrode has with the transformer, it must carry the 14,000 V ac from the transformer to the end of the electrodes. If the copper contacts, or the wires used in the direct connection type are worn or broken, they will not be able to perform their intended function.

After you have made the examination of the condition of the electrodes, the condition that is causing the unit not to fire must be corrected. Let's take each condition and I will show you how to correct it.

We will assume that only one of the conditions exists for this exercise. If more than one of these conditions exists, you will want to replace the electrodes, but let's take them one at a time here.

If the electrodes are worn, with a rounded look to them, you will need to dress up the tips. To do this, you will need to remove the electrode from the gun assembly. Loosen the screw located in the center of the gun assembly that holds the electrodes in place. If you have wires that connect the electrode to the transformer, you will find this easier to remove the wires before you slide them out of the holder. Once you have them removed, dress up the tips with a file so that you have a good point on them. Try to follow the original angle of the electrodes as much as possible. Once you have a good point on the electrodes, place them back in the holder, and secure the holding screw finger tight. You may have to use a screw driver to tighten them just enough to hold them in place so that they can be moved with some resistance on them.

You will now have to set the gap on the electrodes as this is critical to the proper operation of the unit. If you do not have an AF11 gauge, set the gap as follows:

1. Set the electrode gap at $1/8$ in.

2. Set the distance from the center of the nozzle to the tip of the electrodes at $1/8$ in.

3. Set the distance up from the center of the nozzle to the tips at $1/4$ in.

Figure 18.6 shows how to set the gaps with the AF11 gauge, but is a good reference for the manual settings as well.

If you noticed cracks in the electrodes, you will have to replace the electrodes. You may have to wipe the electrode ceramic coating off with a cloth to see the cracks. They will be very small in most cases and could be hard to see. The general rule is that if you suspect that you see cracks, replace the electrodes. These are not expensive items, and you will only make the unit operate better by replacing them anyway. You will want to remove the electrodes as described above so that you can match the diameter and length of the new electrodes. Once you have the new electrodes, install them and gap them as described above. The only difference is that you will have to remove the connections from the old electrodes to install them on the new ones. At this point, you should examine the connectors for the conditions that are described in item 3. If you notice any signs of wear on the connectors, replace them. To do this, either unsnap the direct connection wires, or remove the nut and washer that holds the copper connections in place and replace them. Once this is done, and the gun assembly is back in place, and all of the connections are made, you are ready to check the operation of the unit.

With the transformer closed, or the access door back in place, turn the power to the unit back on. You may have to press the reset button to get the unit to start. Once the burner motor is running, the unit should fire. If the unit fails to fire, you will need to double check the items listed to make sure that you have all of the settings done properly.

Once the unit is running, turn the power off to the unit. If there is an oil buildup in the fire pot, you will need to allow this to burn off before you can check the cycle. Turn the thermostat down so that the unit is not calling for heat, and turn the power back on to the unit. This will allow the blower to come on and not allow the unit to overheat. Once all of the oil has burned off, turn the thermostat back up and check the operation of the unit to make sure that the repairs have been made successfully.

There are many other problems that can occur with the oil forced air heating system that will cause the unit to not operate properly. The majority of these will involve the burner assembly and fuel supply. Others may be caused by the blower assembly. We will examine the other causes here.

Burner Starts and Fires, but Locks Out on Safety

If you have a unit that fires and then locks out on safety, you will need to examine the controls and circuits that can cause this to happen. These would be:

1. Poor fire (nozzle).

2. Flame detector, or stack control.

3. Primary control.

We will need to perform some tests of these items to find out which is causing the problem.

Remove the cover to the cad cell. You will need to jumper the leads on the cad cell to "take it out of the circuit." What I mean by this is that you will be telling the unit that the cad cell is not there. This will allow you to see if this is the problem.

With the cad cell jumped, start the unit, you may need to press the reset to get the unit to fire. Once the unit is running, check to see if the unit locks out. If the unit continues to run, the problem is in the cad cell. There are several areas of the cell that you will need to check.

1. Check to make sure that the face of the cell is not covered with dirt or oil. If the face of the cell is dirty or covered with oil, clean the cell. Any amount of dirt or oil will cause the cell to not see the flame and cause the system to lock out.

2. Check to make sure that the reading of the cell is less than 15,000 ohms. If the reading on the cell is more than 15,000 ohms, you will have to replace the cell as it is faulty.

3. Check the condition of the wires. Are the wires in good shape? If you find any wires that are not in good shape, you will need to repair or replace the wires or holder. Poor wires will cause the cell not to prove properly.

If the cell checks out properly, check the condition of the flame. A poor flame will cause a good cell to not be able to prove the flame and may cause the system to lock out on safety. Check the following conditions of the flame:

1. UNBALANCED FIRE

If you have a fire that is unbalanced, where the flame is not the same all around, you will need to replace the nozzle. This has been described in many areas of this chapter. Refer to that section if the nozzle needs replacing.

2. LEAN SHORT FIRE

This condition is caused by too much air for combustion. Figure 18.9 shows the location of the air control on the burner. Adjust the air flow to clean up the flame.

3. LONG DIRTY FLAME

This condition is caused by too little combustion air. Again, refer to Fig. 18.9 for the location of the combustion air control on the burner and adjust the air to get the proper flame.

If you have checked for all of the above conditions, and have not located the problem, then the problem is in the primary control. You will need to replace this control to resolve the problem.

Once you have corrected the lock out problem, run the unit for a complete cycle to make sure that the problem has been resolved.

Burner Starts, Fires, but Loses Flame and Locks Out on Safety

In some ways, this is the same condition that I just described in the last condition, with one exception. This can also be caused by the fuel supply. As most of the causes will be the same as described in the section on burner fires but locks out, I will not cover the areas for poor fire and cad cell. I will simply add the checks that need to be done on the fuel source.

If the burner loses flame, and the cause cannot be traced to the cad cell of the fire, you will need to check the following:

If the burner starts, loses flame, but does not lock out, the problem is with the fuel supply. You will need to check these items.

1. PUMP LOSES PRIME

If the pump loses its prime, there is an air slug in the line. You will need to bleed the pump. To do this, refer to Fig. 18.9 to locate the bleeder valve on the burner. Start the unit (you may have to press

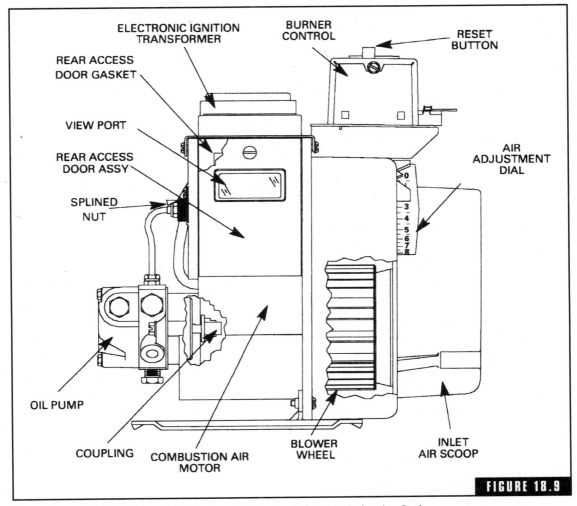

Burner assembly with view of air adjustment. (*Courtesy of Lennox Industries, Inc.*)

the reset button) and open the bleeder valve. You will want to hold a jar under the bleeder to catch the oil. Continue to bleed the pump until you have a clear flow of oil. Close the bleeder while the pump is running so that you do not get more air into the system.

2. PUMP LOSES PRIME—AIR LEAK IN SYSTEM

If you have been bleeding the system, and cannot get a good flow of oil, you could have an air leak in the fuel line. Check all of the connections to make sure that you have the connections tight. Once you have located the problem, you will need to bleed the pump in the same manner as described in item #1.

3. WATER SLUG IN THE LINE

If you have a water slug in the line, you will need to check the tank to see how much water is in the tank. Water and fuel do not mix, and water will not burn. If there is more than 1 in of water in the tank, you will need to have the homeowner call the fuel company to have the tank pumped out to remove the water before you can repair the problem. Once the tank has been pumped and filled, you will want to replace the tank filter, and bleed the system until you get a good clean oil flow. Once you have a good oil flow, close the bleeder with the pump running so that you do not get any air into the line.

4. PARTIALLY PLUGGED NOZZLE

If you found a problem with water in the oil, you could experience a problem with the strainer being plugged on the nozzle. If this is the case, you will need to replace the nozzle as described in this section.

5. RESTRICTION IN THE LINE

You could have a restriction in the oil line. You will have to check the line to see where the restriction is and remove the restriction. This will involve "breaking" the line as close to the restriction as possible. This will mean that air will enter the system so that you will have to bleed the system prior to starting.

In all of these cases, once you have corrected the problem, run the system to make sure that the problem has been resolved.

Burner Starts and Fires, but Short Cycles

If the problem is that the system seems to be short cycling, and is not producing the proper amount of heat, check these items to correct the problem.

Thermostat

1. HEAT ANTICIPATOR SET TOO LOW

The heat anticipator is located on the bottom of the thermostat. It is marked in amps and must be properly set to keep the system operating properly. Change the setting on the heat anticipator to the proper setting if it has been changed. Check with the manufacturer of the thermostat for the proper setting.

2. VIBRATION AT THE THERMOSTAT

This is not a common problem, but it will cause a problem with the operation of the heating system. You will need to locate the source of the vibration and correct this problem.

3. THERMOSTAT LOCATED IN A WARM AIR DRAFT

Check to see if the thermostat is located in an area where the warm air from the outlet in the room is blowing directly on the thermostat. This will cause the unit to short cycle as the thermostat is being heated instead of the room. As the thermostat is the switch that controls the heating unit, it must be located in an area that will not be affected by draft. You need to make sure that the thermostat is not located in direct sunlight. This will keep the unit from operating as well. In any of these cases, you will need to change the location of the thermostat so that it will operate properly.

Blower

Many of the problems with short cycling can be traced to the unit blower. If the blower or any of the components are not working properly, you will not get the proper air flow from the unit. Check these items on the blower.

1. DIRTY AIR FILTER

If the filter is dirty or plugged, the amount of air that can be pulled into the blower to be heated will be restricted. Replace the filter to increase air flow.

2. BLOWER RUNNING TOO SLOW

If you have a multispeed blower that is direct drive, change the speed of the blower by changing the wiring. On the newer units, this is done on the BCSS board as seen in Fig. 18.10.

If you have an older direct drive multispeed blower, change the wires to the next higher speed wire. Refer to the wiring diagram located on the blower door of most units.

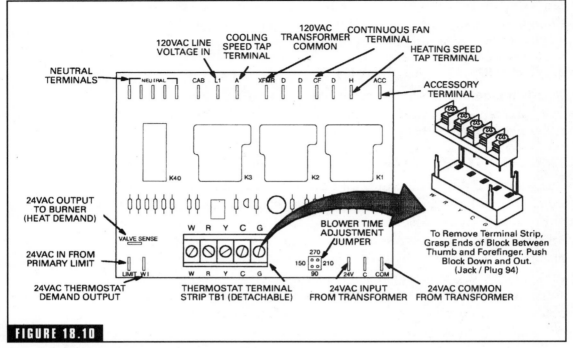

FIGURE 18.10

BCC2 blower board for adjusting blower speed. (*Courtesy of Lennox Industries, Inc.*)

If the unit has a belt drive blower, loosen the set screw on the front end of the blower motor pulley. Turn the pulley counter clockwise to increase the speed. Tighten the set screw once the speed has been adjusted.

In any of these cases, you should check the temperature rise to make sure that you have the speed adjusted properly. Figure 18.11 shows the location of the thermometers for checking this rise.

3. BLOWER MOTOR SEIZED

If the blower motor is seized, you will need to replace the blower motor. If you have a direct drive motor, turn the power off to the unit. Disconnect the wiring from either the BCSS control, or wire cabinet. Loosen the set screw that holds the cage to the motor shaft. Remove the bolts that hold the motor to the blower housing. Slide the motor out of the housing. Loosen the bolts that hold the mounting bracket to the motor and remove. Replace the motor with one of the same voltage and RPM rating. If this is a multispeed motor, be sure to replace with another multispeed motor. Reverse the procedure to replace motor to the housing.

If you have a belt drive unit, turn the power off to the unit. Remove the belt from the motor pulley. Loosen the set screw that holds the pulley to the motor shaft and remove. Loosen the mounting brackets that hold the motor to the motor bracket and remove. Loosen the screws that hold the wire cover in place and on the back of the motor and remove the cover. Remove any conduit that is connected to the motor. Loosen the nuts that hold the wires in place and remove the wires. Replace the motor with one of the same voltage and RPM rating. Replace the motor by reversing this procedure.

Once you have the motor replaced and mounted to the bracket, slide the pulley into place. Install the belt on both pulleys. Slide the pulley on the motor until the belt is in a straight line with the blower pulley. Once this is done, tighten the motor pulley into

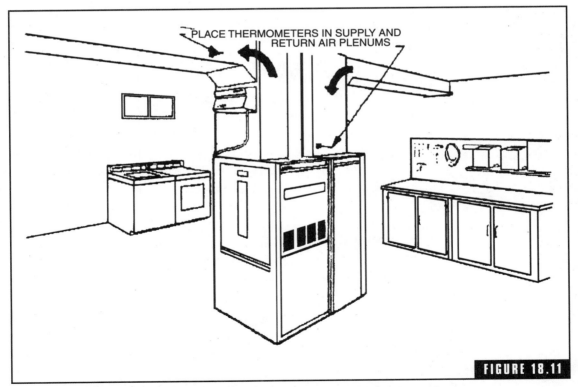

FIGURE 18.11

Placement of thermometers for temperature rise. (*Courtesy of Lennox Industries, Inc.*)

place. Run the blower to make sure that the belt tracks straight. Adjust as needed.

4. BLOWER WHEEL DIRTY

If the blower wheel is dirty, clean the fins. Dirty fins on the blower will restrict the amount of air flow from the blower.

5. PROBLEMS WITH THE MAIN POWER SUPPLY

If you have checked all of these items, and the problem still exists, the problem could be caused by the main power supply. If you suspect this to be the case, call the local power company and have them check this problem for you.

Burner Will Not Shut Off

If you run into a problem where the burner will not shut off, it can be traced back to either the thermostat or the primary control. In most thermostat problems, you will want to replace the thermostat. Here are some of the causes and corrections.

1. SHORTED OR WELDED THERMOSTAT CONTACTS

If you have an older thermostat with contact points, you could have a problem with the contacts becoming welded together not allowing the circuit to open. If this is the case, you will have to replace the thermostat. As this is a straightforward procedure, we will not go into the details of replacing the thermostat. Just be sure to check the new thermostat for level. The thermostat must be level to operate properly.

2. POINTS STUCK

This is different from the points being welded. In this case, you can clean the contacts to correct the problem. Run a clean piece of paper between the points to clean the points. Run the unit to make sure that you have solved the problem.

3. SHORTED THERMOSTAT WIRES

If you have shorted wires, you will need to repair the wires. If this is at the thermostat, there should be enough wire in the wall that can be pulled out until you get to a good section of wire. If the problem is in the wires in the wall, you will need to have an installer come out and replace the wire in the wall in most cases. To check for this, disconnect the wires at the thermostat. If the unit continues to run, this could be the cause. Before you call the installer, also check the primary control as this can also be the problem. Disconnect the thermostat wire at the primary control. If the burner continues to run after the thermostat is disconnected at the primary control, this is the problem, not the thermostat, and you will need to replace the primary control.

If the unit does shut off when you disconnect the thermostat at the primary control, then the problem is the thermostat wire.

As I mentioned at the start of this chapter, the oil forced air heating system is the most challenging unit to troubleshoot and repair of all the units in this book. I have made every effort to cover the most common, and some of the not so common, causes of problems with this type of unit. I am including troubleshooting chart 18.1 to help you better understand the procedure used to troubleshoot these units.

Because of the amount of information that is included in this chapter, you may want to go back over any of the information that you did not understand, or that you need more study time on. You may also want to refer to Chap. 17 for the location of any parts that are on the unit that was not covered in this chapter. I have only included those illustrations that I felt were needed in this chapter to better illustrate the points needed here.

TROUBLE	SOURCE	PROCEDURE	CAUSES	CORRECTION
BURNER FAILS TO START	THERMOSTAT	Check thermostat settings.	Thermostat in **OFF** or **COOL**.	Switch to **HEAT**.
			Thermostat set too low.	Turn thermostat to higher temp.
	SAFETY OVER LOADS	Check burner motor, primary safety control, & auxiliary limit switch.	Burner motor overload tripped.	Push pump motor reset button.
			Primary control tripped on safety.	Reset primary control.
			Auxiliary limit switch tripped on safety.	Reset auxiliary limit.
	POWER	Check furnace disconnect switch & main disconnect.	Open Switch.	Close switch.
			Blown fuse or tripped circuit breaker.	Replace fuse or reset circuit breaker.
	THERMOSTAT	Touch jumper wire across thermostat terminals on primary control. If burner starts, then fault is in thermostat circuit.	Broken or loose thermostat wires.	Repair or replace wires.
			Loose thermostat screw connection.	Tighten connection.
			Dirty thermostat contacts.	Clean contacts.
			Thermostat not level.	Level thermostat.
			Faulty thermostat.	Replace thermostat.
	CAD CELL	Disconnect flame detector wires at primary control. If burner starts, fault is in detector circuit.	Flame detector leads shorted.	Separate leads.
			Flame detector exposed to light.	Seal off false source of light.
			Short circuit in flame detector.	Replace detector.
	PRIMARY CONTROL	Place trouble light between the black and white leads. No light indicates no power to control.	Primary or auxiliary control switch open.	Check adjustment. Set to maximum setting.
				Jumper terminals; if burner starts, switch is faulty, replace control.
			Open circuit between disconnect switch and limit control.	Trace wiring and repair or replace.
			Low line voltage or power failure.	Call power company.
		Place trouble light between the orange and white leads. No light indicates faulty control.	Defective internal control circuit.	Replace control.
	BURNER	Place trouble light between the black & white leads to burner motor. No light indicates no power to motor.	Blown fuse.	Replace fuse.
		Place trouble light between the black & white leads to burner motor. Light indicates power to motor & burner fault.	Binding burner blower wheel.	Turn off power & rotate blower wheel by hand. If seized, free wheel or relpace fuel pump.
			Seized fuel pump.	
			Defective burner motor.	Replace motor.
BURNER STARTS BUT NO FLAME IS ESTABLISHED	OIL SUPPLY	Check tank gauge or use dip stick.	No oil in tank.	Fill tank.
		Coat dip stick with litmus paste & insert to bottom of tank.	Water in oil tank.	If water depth exceeds 1" (25.4mm), pump or drain out water.
		Listen for pump whine.	Tank shut-off valve closed.	Open valve.
	OIL FILTERS & OIL LINE	Listen for pump whine.	Oil line filter plugged.	Replace filter cartridge.
			Kinks or restriction in oil line.	Repair or replace oil line.
			Plugged fuel pump strainer.	Clean strainer or replace pump.
		Open bleed valve or gauge port. Start burner. No oil or milky oil indicates loss of prime.	Air leak in oil supply line.	Locate and correct leak.
				Tighten all connections.
	OIL PUMP	Install pressure gauge on pump & read pressure. Should not be less than 140 psi (965.3 kPa) or 100 psi (689.5 kPa) for O23Q2-70 units.	Pump partially or completely frozen —No pressure and motor locks out on overload.	Replace pump.
			Coupling disengaged or broken —No pressure.	Re-engage or replace coupling.
			Fuel pressure too low.	Adjust to 100 psi (689.5 kPa) or 140 psi 965.3 kPa).
	NOZZLE	Disconnect ignition leads. Observe oil spray (gun assembly must be removed from unit). Inspect nozzle for plugged orifice or carbon build-up around orifice.	Nozzle orifice plugged.	Replace nozzle with same size, spray angle and spray type.
			Nozzle strainer plugged.	
			Poor or off center spray.	

TROUBLESHOOTING CHART 18.1   Oil forced air heating.

222

TROUBLE	SOURCE	PROCEDURE		CAUSES	CORRECTION
CONTINUED BURNER STARTS BUT NO FLAME IS ESTABLISHED	IGNITION ELECTRODES	Remove gun assembly and inspect electrodes and leads.		Fouled or shorted electrodes.	Clean electrode leads.
				Dirty electrodes and leads.	
				Eroded electrode tips.	Dress-up electrode tips & reset gap to 1/8" (3.2mm) and correctly position tips using the AFII multipurpose gauge T-500.
				Improper electrode gap spacing.	
				Improper position of electrode tips.	
				Cracked or chipped insulators.	Replace electrode.
				Cracked or burned lead insulators.	Replace electrode Leads.
	IGNITION TRANSFORMER	Connect ignition leads to transformer. Start burner and observe spark. Check line voltage to transformer primary.		Low line voltage.	Check voltage at power source. Correct cause of voltage drop or call power company.
				No spark or weak spark.	Properly ground transformer case.
	BURNER MOTOR	Motor does not come up to speed and trips out on overload. Turn off power and rotate blower wheel by hand to check for binding or excessive drag.		Low line voltage.	Check voltage at power source. Correct cause of voltage drop or call power company.
				Pump or blower overloading motor.	Correct cause of overloading.
				Faulty motor.	Replace motor.
BURNER STARTS & FIRES BUT LOCKS OUT ON SAFETY	POOR FIRE	After burner fires, immediately jumper across flame detector terminals at primary control.	If burner continues to run, fault may be due to poor fire. Inspect fire.	Unbalanced fire.	Replace nozzle.
				Too much air – lean short fire.	Reduce combustion air - Check combustion.
				Too little air – long dirty fire.	Increase combustion air-Check combustion.
				Excessive draft.	Adjust Barometric damper for correct draft.
				Too little draft or restriction.	Correct draft or remove restriction.
	FLAME DETECTOR		If fire is good, fault is in the flame detector. Check detector circuit.	Dirty cad cell face.	Clean cad cell face.
				Faulty cad cell – exceeds 15000 ohms.	Replace cad cell.
				Loose or defective cad cell wires.	Secure connections or replace cad cell holder and wire leads.
	PRIMARY CONTROL		If burner locks out on safety, fault is in primary control.	Primary control circuit defective.	Replace primary control.
BURNER STARTS, FIRES BUT LOOSES FLAME & LOCKS OUT ON SAFETY	POOR FIRE	After burner fires, immediately jumper across flame detector terminals at primary control.	If burner continues to run (does not lock out on safety), fault may be poor fire (marginal). Inspect fire.	Unbalanced fire.	Replace nozzle.
				Too much air – lean short fire.	Reduce combustion air – check combustion.
				Too little air – long dirty fire.	Increase combustion air – check combustion.
				Excessive draft.	Adjust barometric damper for correct draft.
				Too little draft or restriction.	Correct draft or remove restriction.
	FLAME DETECTOR		If fire is good fault is in the flame detector. Check detector circuit.	Dirty cad cell face.	Clean cad cell face.
				Faulty cad cell – exceeds 15000 ohms.	Replace cad cell.
				Loose or defective cad cell wires.	Secure connections or replace cad cell holder and wire leads.
	OIL SUPPLY		If burner loses flame (does not lock out on safety), fault is in fuel system.	Pump loses prime – air slug.	Prime pump at bleed port.
				Pump loses prime – air leak in supply line.	Check supply line for loose connections and tighten fittings.
				Water slug in line.	Check oil tank for water (over 1" [25.4mm]) pump or drain out water.
				Partially plugged nozzle or nozzle strainer.	Replace nozzle.
		Listen for pump whine.		Restriction in oil line.	Clear restriction.
				Plugged fuel pump strainer.	Clean strainer or replace pump.
				Cold oil – outdoor tank.	Change to number 1 oil.

TROUBLESHOOTING CHART 18.1 (*continued*)   Oil forced air heating.

TROUBLE	SOURCE	PROCEDURE	CAUSES	CORRECTION
BURNER STARTS AND FIRES BUT SHORT CYCLES (TOO LITTLE HEAT)	THERMOSTAT	Check thermostat.	Heat anticipator set too low.	Correct heat anticipator setting.
			Vibration at thermostat.	Correct source of vibration.
			Thermostat in warm air draft.	Shield thermostat from draft or relocate.
	LIMIT CONTROL	Connect voltmeter between line voltage connections to primary control (black & white leads). If burner cycles due to power interruption, it's cycling off limit.	Dirty furnace air filters.	Clean or replace filter.
			Blower running too slow.	Increase blower speed to maintain proper temp. rise.
			Blower motor seized or burned out.	Replace motor.
			Blower wheel dirty.	Clean blower wheel.
			Blower wheel in backwards.	Reverse blower wheel.
			Wrong motor rotation.	Replace with properly rotating wheel.
			Restrictions in return or supply air system.	Correct cause of restriction.
			Adjustable limit control set too low.	Reset limt to maximum stop setting.
	POWER	If voltage fluctuates, fault is in the power source. Recheck voltage at power source.	Loose wiring connection.	Locate and secure connection.
			Low or fluctuating line voltage.	Call power company.
BURNER RUNS CONTINUOUSLY (TOO MUCH HEAT)	THERMOSTAT	Disconnect thermostat wires at primary control.	If burner turns off, fault is in thermostat circuit.	Shorted or welded thermostat contacts.	Repair or replace thermostat.
			Stuck thermostat bimetal.	Clear obstruction or replace thermostat.
			Thermostat not level.	Level thermostat.
			Shorted thermostat wires.	Repair short or replace wires.
			Thermostat out of calibration.	Replace thermostat.
			Thermostat in cold draft.	Correct draft or relocate thermostat.
	PRIMARY CONTROL		If burner does not turn off, fault is in primary control.	Defective primary control.	Replace defective primary control.
BURNER RUNS CONTINUOUSLY (TOO LITTLE HEAT)	COMBUSTION	Check burner combustion for CO_2, stack temperature & smoke.	Low CO_2 less than 10%.	Too much combustion air.	Reduce combustion air.
			Air leaks into heat exchanger around inspection door, etc.	Correct cause of air leak.
			Excessive draft.	Adjust barometeric damper for correct draft.
			Incorrect burner head adjustment.	Correct burner head setting.
		High smoke reading more than a trace.	Dirty or plugged heat exchanger.	Clean heat exchanger.
				Readjust burner.
			Insufficient draft.	Increase draft.
			Incorrect burner head adjustment.	Correct burner setting.
			Too little combustion air.	Increase combustion air.
		High stack temperature more than 550°F (288°C) Net.	Too little blower air.	Increase blower speed to maintain proper temp. rise.
			Dirty or plugged heat exchanger.	Clean heat exchanger.
			Dirty blower wheel.	Clean blower wheel.
			Dirty furnace air filters.	Clean or replace filter.
			Restricted or closed registers or dampers.	Readjust registers or dampers.
	OIL PRESSURE	Inspect fire and check oil pressure.	Partially plugged or defective nozzle.	Replace nozzle.
			Oil pressure too low, less than 140 (965.3 kPa) psi or 100 psi (689.5 kPa) for O23Q2–70 units.	Increase oil pressure to 140 psi (965.3 kPa) or 100psi (689.5 kPa) for O23Q2–70 units.

TROUBLESHOOTING CHART 18.1 (*continued*)   **Oil forced air heating.**

Is Electric Forced Air Heat Right for You?

Electric forced air heat, also known as *resistance heat,* is a very energy-efficient way to heat your home in some parts of the country. But is it the right heat source for you?

To answer this question, you must first look at what part of the country you live in. If you live in a part of the country that has very cold winters, this may not be the correct heat source for you. However, if you live in a part of the country where the winters are mild and the cost of electricity is low, this could be the right choice for you. One item that you must remember when considering the use of electricity to heat your home is that this is not a natural resource like gas or oil. It does, however, require one of the natural resources to make the electricity. This can cause the cost of a kilowatt-hour to be quite high in some parts of the country. In those states where the cost of a kilowatthour is high, electricity may not be the best choice to heat your home.

Electricity is a versatile but precious resource. Because it is needed for refined power equipment such as medical equipment and computers, when it is used for less refined needs, such as heating a home, it must be used as efficiently as possible. If you use electricity to heat your home, there are ways to make your home as energy efficient as possible. This will not only help to preserve this resource, but it will help to save money and to reduce your energy consumption as well.

Energy Savings Measures

Insulation

To keep your heating costs reasonable, homes using electric forced air heating systems should be very well insulated. The insulation's ability to control the amount of heat flow is measured in R value (R stands for thermal resistance). The higher the R value, the better the insulation restricts heat flow.

You must make sure that you have an adequate amount of insulation in the walls, ceilings, and under the floor of your home. You also must make sure that the insulation was installed properly so that there are no gaps or voids. Any gaps or voids in the insulation will allow air convection or air leakage that will greatly reduce the R value of the insulation in the home.

Windows

Instead of R values, windows are usually rated by their heat transfer coefficient, or U value. The lower the U value, the better is the window's thermal resistance, or resistance to heat loss.

While energy-efficient windows are important in any home, electrically heated homes especially should have windows with U values of less than 0.40. Advances in window design incorporate multiglazing layers, heat-reflective coatings, or gas fillings to reduce the U value to less than 0.25. If you own a home that is electrically heated, even with energy-efficient windows, installing storm windows can help to reduce your energy costs if you live in a cold climate where the cost of electricity is high.

Reducing Air Leaks

Even if you have a new home that was built to high energy standards (sometimes called "Super Good Cents"), air leaks can occur. To help prevent these energy-robbing leaks, there are a few steps that you can take.

Homeowners need to make sure that they have proper caulking around all the windows in the home. There should be no cracks or voids in the caulk that would allow air to escape from the home. Homeowners also should check the weather stripping around all the exterior doors in the home. Make sure that the weather stripping is in good shape and that there are no cracks or gaps in it.

Insulated gaskets should be installed on all electrical outlets. These are very inexpensive and are installed behind the covers on the wall. You would be amazed at the amount of cold air that can enter a home from behind these covers.

If you are having a new home built and are planning on having electric forced air heating installed, ask your contractor if he or she is going to install an air infiltration barrier around the home. The contractor also should seal all joints and penetrations in the home.

If you are doing a new construction, you should be aware of some areas that will make sealing most, if not all leaks, almost impossible. You should avoid such things as

1. Complicated floor plans
2. Irregular roof lines
3. Protruding windows
4. Cathedral ceilings
5. Fireplaces
6. Recessed lighting fixtures

As a result of having some or all of these features in your home, often you will have higher heating costs due to excessive air leakage.

Forced Air Heating Ductwork

A forced air heating system's ductwork also influences residential air leakage. Homes with forced air heating systems can have a higher rate of air leakage than those with electric baseboard heating due to air leakage from the heating system's ductwork. Heat is lost through leaky or uninsulated ducts. Joints between the sections of the ducts, between the ducts and the registers, and between the ducts and the furnace can lose as much as 30 percent of the heating efficiency.

Leaking ductwork can cause a positive and negative pressure that often increases air leakage through floors, exterior walls, and ceilings. Reducing or eliminating air leakage will make your home more comfortable and energy efficient and will help to reduce your energy costs as well.

Filters

Another area to remember to check when talking about the efficiency of your heating system is the filter. The filter is designed to keep the heat exchanger, blower, and ductwork clean. If the filter becomes dirty and cannot allow the proper amount of air to pass through to the blower, you will not get the proper amount of airflow from your heating system. This will then cause the heating system to have to operate for a longer period of time to keep your home at the required temperature. This also will cause your energy costs to rise. You should remember to replace your filter every 2 months to help reduce your energy costs.

Conclusion

As you can see, there are many factors to consider when installing or operating an electric forced air heating system to get the most for your energy dollar. Electric heating is considered 100 percent efficient because it converts 100 percent of every kilowatt of electricity to heat. The cost of a kilowatt of electricity can determine if this is the right choice of heating system for your home. You also must consider the age of the home and the condition of the windows and insulation. Most new homes are designed to be as energy efficient as possible. In some parts of the country, electric heat is the only option. In such a situation, if you follow the guidelines in this chapter, you should save money on your heating costs.

One other way to save money on your electric heating costs is to install an add-on heat pump. Most electric forced air heating systems come equipped to handle an add-on heat pump. These units come in many different styles and energy ratings depending on what part of the country you live in. Heat pumps will be covered in detail in Section 5 of this book.

Controls for an Electric Forced Air Heating System

The electric forced air heating system has fewer controls than the other systems covered in this book. However, the controls of an electric forced air heating system perform the same basic functions.

Since this is an electric forced air heating system, the use of electricity and the voltages used are critical to the proper operation of the unit. The units will use voltages of between 208 and 575 V ac. This is possible by the use of step-up transformers that can convert the line voltage to a much higher voltage needed to control the circuits. We will now examine these circuits and how they work together. Figure 20.1 shows the basic parts layout for an electric downflow forced air heating system with a cooling coil installed.

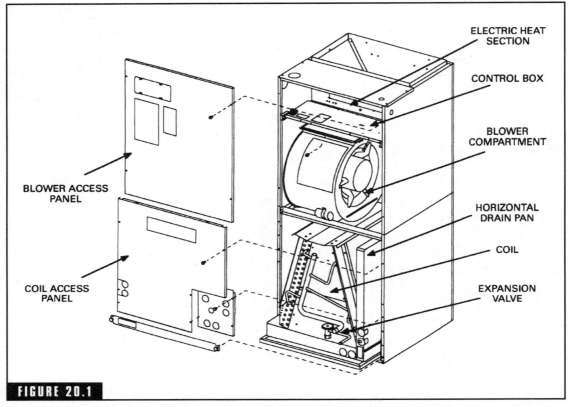

FIGURE 20.1

Components of an electric forced air heating system. (*Courtesy of Lennox Industries, Inc.*)

Thermostat

The thermostat used in an electric forced air heating system is the same as that used in the other types of forced air heating systems covered so far in this book. It is a 24-V ac thermostat. With an electric forced air heating system, this thermostat will, in most cases, be equipped with levers to set the unit for heating/cooling as well as a blower lever to allow for manual or automatic operation of the blower. During the heating season, these controls are set for heating and automatic operation of the blower to allow for the unit to control the operation of the blower.

Control Box

Figure 20.2 shows the control box. This is where the line voltage and electrical connections are made. This is also where the electric heating elements are housed.

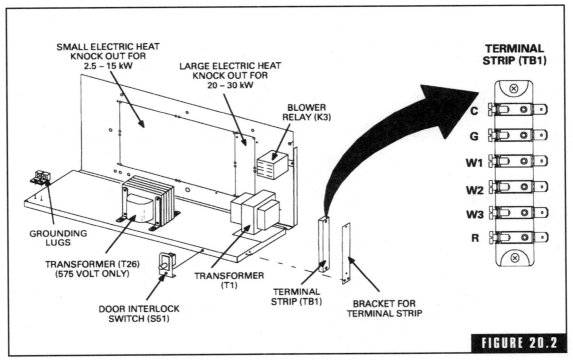

Control box. (*Courtesy of Lennox Industries, Inc.*)

Door Interlock Switch

All units that operate at the 460 to 575 V ac will have a door interlock switch wired in series with the terminal strip. This switch operates on 125 V ac and will shut the unit down when the blower door is opened. This switch is located on the control box.

Terminal Strip

The terminal strip (see Fig. 20.2) is where the thermostat connections are made. If the unit is going to be wired for central air or a heat pump, these thermostat connections also would be made here. Any outdoor low-voltage connections that are to be made can be spliced and connected with wire nuts in the control box as well.

Transformer

The 24-V ac transformer is located in the control box (see Fig. 20.2). This supplies voltage to the indoor and outdoor low-voltage circuits. These

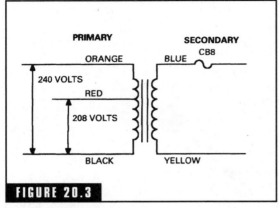

FIGURE 20.3

208/240-V transformer.

transformers are rated at 70 VA. All 208/240-V ac single-phase transformers use two primary voltage taps, as shown in Fig. 20.3.

Circuit Breakers

All 208/240-V ac transformers also are equipped with an internal secondary overcurrent protection. Each transformer uses a circuit breaker located on the transformer. The circuit breaker is wired in series with the blue secondary voltage wire and is rated at 3.5 A.

Autotransformer

On larger units, a 575- to 460-V ac step-down transformer is used. This transformer is mounted in the control box (see Fig. 20.2). This transformer comes with a 575-V ac electric heater and is connected to the unit via jacks/plugs. This transformer supplies 460 V ac to the blower motor on these units.

Transformer Fuses

On the units that use 575-V ac transformers, these are protected by two line fuses. Both these fuses are rated at 600 V and 3.2 A.

Blower Relay

On all units that operate on less than 460 V ac, a DPDT relay is used to energize the blower motor. This relay is located in the control box (see Fig. 20.2). The relay coil is energized by the heating demand from the thermostat. A set of NC contacts are used to allow the electric heat relay to energize and the blower to operate in the heating speed mode. Figure 20.4 shows the wire diagram for the 208/240-V ac electrical system.

Blower Motor

Figure 20.5 shows the blower motor and run capacitor. On all units that operate at 208/240 V ac, the motors are single-phase direct-drive

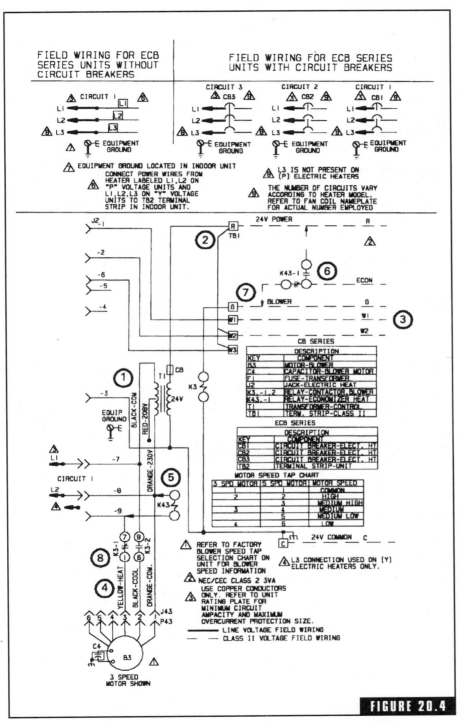

Wiring diagram of a 208/240-V ac system. (*Courtesy of Lennox Industries, Inc.*)

units that use a run capacitor. All these motors are equipped with multiple-speed taps. Typically, the high-speed tap is energized during normal operation. The table in Fig. 20.6 shows the horsepower rating and capacitor rating for the CB29M and CB30M electric forced air heating units supplied by Lennox International.

460-V Motor Winding

A third tap (blue) on 460-V ac motors is used for internal wiring during low-speed operation and must not be connected to the line voltage. During low-speed (yellow tap) operation, the high-speed (black) tap is disconnected from the line voltage and is connected to the blue internal wiring tap. This is done by the blower relay.

Blower Motor Capacitor

All units that operate at 208/240 V ac use single-phase direct-drive motors with a run capacitor. The run capacitor is mounted on the blower housing, as seen in Fig. 20.5. The rating of the capacitor is listed in the chart in Fig. 20.6.

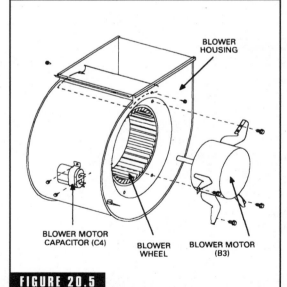

BLOWER
HOUSING

BLOWER MOTOR
CAPACITOR (C4)

BLOWER
WHEEL

BLOWER MOTOR
(B3)

FIGURE 20.5

Blower motor and assembly. (*Courtesy of Lennox Industries, Inc.*)

Electric Heat Components

The parts arrangements are shown in Figs. 20.7 to 20.11. All electric heating sections consist of components mounted to the electric heat vestibule panel, and the electric heating elements are exposed directly to the airstream. All 208/240-V ac heating units are equipped with circuit breakers, whereas 480- and 575-V ac units are protected by fuses.

Electric Heat Sequencer Relays (208/240-V ac Units)

The electric heat sequencer relays are NO relays with a resistive element for a coil and a bimetallic disk that activates the contacts. The relays are located in the electric heat vestibule and are energized by a 24-V heating demand (see Figs. 20.7 to 20.11). When energized, the internal resistance heats the bimetallic disk, causing the contacts to close. When the relay is deenergized, the disk cools, and the contacts open. The relays energize different stages of heat, as well as the blower. The blower is always first on and last off.

Heat Blower Relay

This is a three-pole double-throw (3PDT) NO contactor located in the electric heat vestibule panel. The contactor is equipped with a 24-V coil and is energized by a 24-V heating demand. The contactor energizes the blower, as well as the heat relay. The blower is always first on and last off.

UNIT	HORSE POWER	CAPACITOR RATING
CB29M-21/26 (P)	1/5 HP	7.5MFD / 370V
CB29M-31 (P)	1/3 HP	5MFD / 370V
CB29M-41 (P)	1/3 HP	5MFD / 370V
CB29M-46 (P/G)	1/2HP	10MFD / 370V
CB29M-51 (P/G)	3/4 HP	10MFD / 370V
CB29M-65 (P/G)	3/4 HP	20MFD / 370V
CB30M-21/26 (P)	1/5 HP	7.5MFD / 370V
CB30M-31 (P)	1/3 HP	15MFD / 370V
CB30M-41 (P)	1/3 HP	15MFD / 370V
CB30M-46 (P)	1/3 HP	20MFD / 370V
CB30M-51 (P/G)	1/3 HP	20MFD / 370V
CB30M-65 (P/G)	1/2 HP	20MFD / 370V

FIGURE 20.6

Horsepower rating chart. (*Courtesy of Lennox Industries, Inc.*)

Electric Heat Contactor (460- and 575-V Units Only)

This contactor is a three-pole double-break (3PDB) NO contactor located on the electric heat vestibule panel. The contactor is equipped with a 24-V coil and is energized by a 24-V heat demand. The contactor energizes the heating elements.

Primary and Secondary Temperature Limits

Both the primary and secondary limits are located in the heating vestibule panel and are exposed directly to the airstream through an opening in the panel. The high limits are SPST NC limits, with the primary limit being an autoreset limit and the secondary limit being a "one time" limit. One-time limits must be replaced when opened. The limits are factory set and are not adjustable.

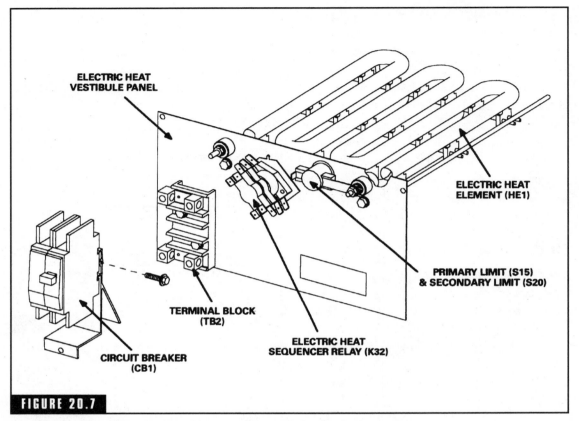

ELECTRIC HEAT
VESTIBULE PANEL

ELECTRIC HEAT
ELEMENT (HE1)

PRIMARY LIMIT (S15)
& SECONDARY LIMIT (S20)

TERMINAL BLOCK
(TB2)

ELECTRIC HEAT
SEQUENCER RELAY (K32)

CIRCUIT BREAKER
(CB1)

FIGURE 20.7

208/230-V ac parts arrangements. (*Courtesy of Lennox Industries, Inc.*)

208/240-V Electric Heat Sections

Each stage of the 208/240-V electric heating unit is protected by a primary and secondary high-temperature limit. Both are located in the same housing. Each stage uses the same style of limits. Both the primary and secondary limits are wired in series with the heating elements. When either of the limits opens, the corresponding heating element is deenergized. All other heating elements stay energized. The primary limit opens at 150°F plus or minus 5°F on a temperature rise and automatically resets at 110°F plus or minus 9°F on the temperature fall. The secondary high-temperature limit opens at 333°F plus or minus 10°F on a temperature rise. If the secondary limit opens, it will need to be replaced.

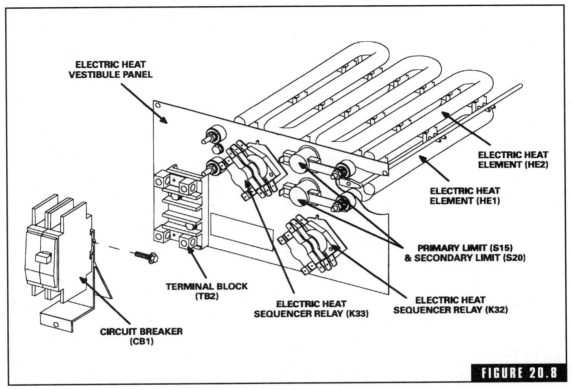

208/230-V ac parts arrangement. (*Courtesy of Lennox Industries, Inc.*)

460- AND 575-V Electric Heating Sections

The 460- and 575-V electric heating sections are protected by three separate high-temperature limits: one primary and two secondary limits. The primary is wired in series with the contactor coil, whereas the secondary limits are wired in series with the heating elements after the contacts. When the primary opens, all heating elements are deenergized. If either of the secondary limits opens, the heating output is cut in half. If both the secondary limits open, all elements are deenergized.

The primary high limit opens at 150°F plus or minus 5°F on a temperature rise and automatically resets at 110°F plus or minus 9°F on a temperature fall. The secondary high limit opens at 300°F plus or minus 10°F on a temperature rise. If the secondary limit opens, it must be replaced.

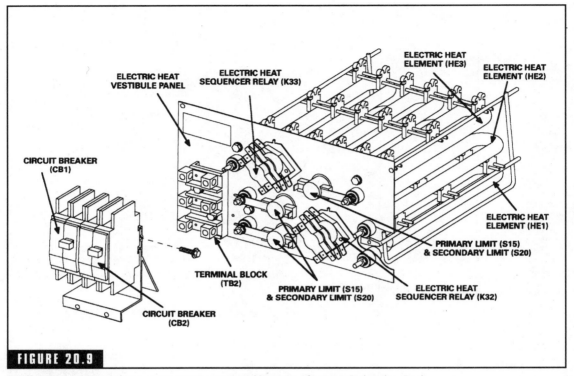

FIGURE 20.9

208/230-V ac three-phase parts arrangements. (*Courtesy of Lennox Industries, Inc.*)

Heating Elements

Heating elements are composed of helix-wound bare nichrome wire exposed to the airstream. The elements are supported by insulators mounted to the wire frame. For single-phase applications, one element is used per stage. Each stage is energized independently by the corresponding relay located on the electric heat vestibule panel. All three-phase heating elements are arranged in a delta pattern. Once energized, heat transfer is instantaneous. High-temperature protection is provided by primary and secondary temperature limits. Figures 20.12 to 20.22 show the electric heat data for single- and three-phase units.

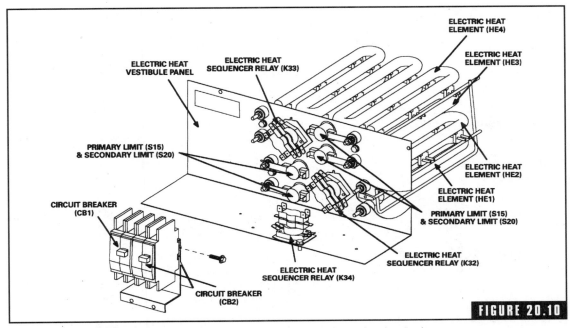

208/230-V ac single-phase parts arrangement. (*Courtesy of Lennox Industries, Inc.*)

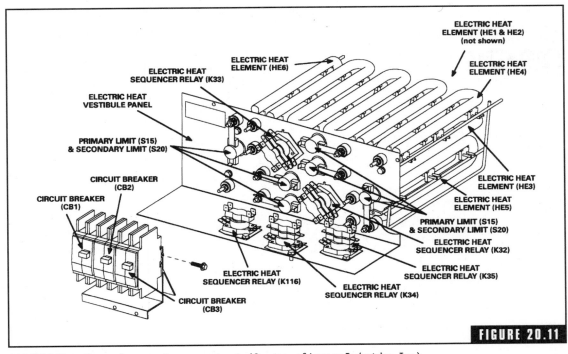

208/230-V ac three-phase parts arrangement. (*Courtesy of Lennox Industries, Inc.*)

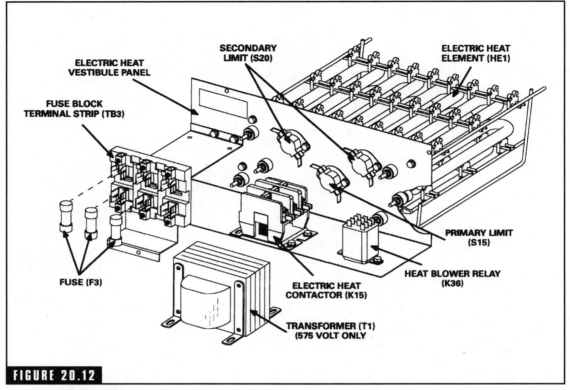

FIGURE 20.12

575-V ac parts arrangement. (*Courtesy of Lennox Industries, Inc.*)

Blower Coil Model Number	Electric Heat kW, Model Number & Shipping Weight		Number of Steps	Volts Input	kW Input	**Btuh Input	•Blower Motor Amps	*Minimum Circuit Ampacity	
								Circuit 1	Circuit 2
				◆ 208/230v–1 ph ◆					
CB29M-21/26	2.5 kW	ECB29-2.5 (28K30) 4 lbs. (2 kg)	1 step	208	1.9	6,400	1.5	13	– – – –
				220	2.1	7,200	1.5	14	– – – –
				230	2.3	7,800	1.5	14	– – – –
				240	2.5	8,500	1.5	15	– – – –
	5 kW	ECB29-5 (28K31) ECB29-5CB (28K32) 4 lbs. (2 kg)	1 step	208	3.8	12,800	1.5	25	– – – –
				220	4.2	14,300	1.5	26	– – – –
				230	4.6	15,700	1.5	27	– – – –
				240	5.0	17,100	1.5	28	– – – –
	8 kW	ECB29-8 (28K33) ECB29-8CB (28K34) 5 lbs. (2 kg)	2 steps	208	6.0	20,500	1.5	38	– – – –
				220	6.7	22,900	1.5	40	– – – –
				230	7.3	25,100	1.5	42	– – – –
				240	8.0	27,300	1.5	44	– – – –
	10 kW	ECB29-10 (28K35) ECB29-10CB (28K36) 5 lbs. (2 kg)	2 steps	208	7.5	25,600	1.5	47	– – – –
				220	8.4	28,700	1.5	50	– – – –
				230	9.2	31,400	1.5	52	– – – –
				240	10.0	34,100	1.5	54	– – – –
CB29M-31	5 kW	ECB29-5 (28K31) ECB29-5CB (28K32) 4 lbs. (2 kg)	1 step	208	3.8	12,800	2.4	26	– – – –
				220	4.2	14,300	2.4	27	– – – –
				230	4.6	15,700	2.4	28	– – – –
				240	5.0	17,100	2.4	29	– – – –
	8 kW	ECB29-8 (28K33) ECB29-8CB (28K34) 5 lbs. (2 kg)	2 steps	208	6.0	20,500	2.4	39	– – – –
				220	6.7	22,900	2.4	41	– – – –
				230	7.3	25,100	2.4	43	– – – –
				240	8.0	27,300	2.4	45	– – – –
	10 kW	ECB29-10 (28K35) ECB29-10CB (28K36) 5 lbs. (2 kg)	2 steps	208	7.5	25,600	2.4	48	– – – –
				220	8.4	28,700	2.4	51	– – – –
				230	9.2	31,400	2.4	53	– – – –
				240	10.0	34,100	2.4	55	– – – –
	12.5 kW	ECB29-12.5CB (28K37) 10 lbs. (5 kg)	3 steps	208	9.4	32,000	2.4	22	38
				220	10.5	35,800	2.4	23	40
				230	11.5	39,200	2.4	24	42
				240	12.5	42,600	2.4	25	44
	15 kW	ECB29-15CB (28K38) 10 lbs. (5 kg)	3 steps	208	11.3	38,400	2.4	26	45
				220	12.6	43,000	2.4	27	48
				230	13.5	47,000	2.4	28	49
				240	15.0	51,200	2.4	29	52

*Refer to National or Canadian Electrical Code manual to determine wire, fuse and disconnect size requirements. Use wires suitable for at least 167°F (75°C).
**Electric heater capacity only — does not include additional blower motor heat capacity.
•Minimum circuit ampacity for blower motor only.

FIGURE 20.13

Electric heat data (1 phase). (*Courtesy of Lennox Industries, Inc.*)

Blower Coil Model Number	Electric Heat kW, Model Number & Shipping Weight		Number of Steps	Volts Input	kW Input	**Btuh Input	●Blower Motor Amps	*Minimum Circuit Ampacity	
								Circuit 1	Circuit 2
◆ 208/230v–1 ph ◆									
CB29M-41	5 kW	ECB29-5 (28K31) ECB29-5CB (28K32) 4 lbs. (2 kg)	1 step	208	3.8	12,800	2.4	26	- - - -
				220	4.2	14,300	2.4	27	- - - -
				230	4.6	15,700	2.4	28	- - - -
				240	5.0	17,100	2.4	29	- - - -
	8 kW	ECB29-8 (28K33) ECB29-8CB (28K34) 5 lbs. (2 kg)	2 steps	208	6.0	20,500	2.4	39	- - - -
				220	6.7	22,900	2.4	41	- - - -
				230	7.3	25,100	2.4	43	- - - -
				240	8.0	27,300	2.4	45	- - - -
	10 kW	ECB29-10 (28K35) ECB29-10CB (28K36) 5 lbs. (2 kg)	2 steps	208	7.5	25,600	2.4	48	- - - -
				220	8.4	28,700	2.4	51	- - - -
				230	9.2	31,400	2.4	53	- - - -
				240	10.0	34,100	2.4	55	- - - -
	12.5 kW	ECB29-12.5CB (28K37) 10 lbs. (5 kg)	3 steps	208	9.4	32,000	2.4	22	38
				220	10.5	35,800	2.4	23	40
				230	11.5	39,200	2.4	24	42
				240	12.5	42,600	2.4	25	44
	15 kW	ECB29-15CB (28K38) 10 lbs. (5 kg)	3 steps	208	11.3	38,400	2.4	26	45
				220	12.6	43,000	2.4	27	48
				230	13.5	47,000	2.4	28	49
				240	15.0	51,200	2.4	29	52
	20 kW	ECB29-20CB (28K39) 14 lbs. (6 kg)	4 steps	208	15.0	51,200	2.4	48	45
				220	16.8	57,300	2.4	51	48
				230	18.4	62,700	2.4	53	49
				240	20.0	68,200	2.4	55	52
CB29M-46	5 kW	ECB29-5 (28K31) ECB29-5CB (28K32) 4 lbs. (2 kg)	1 step	208	3.8	12,800	3.6	27	- - - -
				220	4.2	14,300	3.6	28	- - - -
				230	4.6	15,700	3.6	30	- - - -
				240	5.0	17,100	3.6	31	- - - -
	8 kW	ECB29-8 (28K33) ECB29-8CB (28K34) 5 lbs. (2 kg)	2 steps	208	6.0	20,500	3.6	41	- - - -
				220	6.7	22,900	3.6	43	- - - -
				230	7.3	25,100	3.6	44	- - - -
				240	8.0	27,300	3.6	46	- - - -
	10 kW	ECB29-10 (28K35) ECB29-10CB (28K36) 5 lbs. (2 kg)	2 steps	208	7.5	25,600	3.6	50	- - - -
				220	8.4	28,700	3.6	52	- - - -
				230	9.2	31,400	3.6	55	- - - -
				240	10.0	34,100	3.6	57	- - - -
	12.5 kW	ECB29-12.5CB (28K38) 10 lbs. (5 kg)	3 steps	208	9.4	32,000	3.6	23	38
				220	10.5	35,800	3.6	24	40
				230	11.5	39,200	3.6	25	42
				240	12.5	42,600	3.6	26	44
	15 kW	ECB29-15CB (28K38) 10 lbs. (5 kg)	3 steps	208	11.3	38,400	3.6	27	45
				220	12.6	43,000	3.6	28	48
				230	13.5	47,000	3.6	29	49
				240	15.0	51,200	3.6	31	52
	20 kW	ECB29-20CB (28K39) 14 lbs. (6 kg)	4 steps	208	15.0	51,200	3.6	50	45
				220	16.8	57,300	3.6	52	48
				230	18.4	62,700	3.6	55	49
				240	20.0	68,200	3.6	57	52

*Refer to National or Canadian Electrical Code manual to determine wire, fuse and disconnect size requirements. Use wires suitable for at least 167°F (75°C).

**Electric heater capacity only — does not include additional blower motor heat capacity. ●Minimum circuit ampacity for blower motor only.

FIGURE 20.14

Electric heat data (1 phase). (*Courtesy of Lennox Industries, Inc.*)

Blower Coil Model Number	Electric Heat kW, Model Number & Shipping Weight	Number of Steps	Volts Input	kW Input	**Btuh Input	●Blower Motor Amps	*Minimum Circuit Ampacity		
							Circuit 1	Circuit 2	Circuit 3
			▼ 208/230v–1 ph ▼						
CB29M-51	5 kW — ECB29-5 (28K31) ECB29-5CB (28K32) 4 lbs. (2 kg)	1 step	208	3.8	12,800	3.8	28	- - - -	- - - -
			220	4.2	14,300	3.8	29	- - - -	- - - -
			230	4.6	15,700	3.8	30	- - - -	- - - -
			240	5.0	17,100	3.8	31	- - - -	- - - -
	8 kW — ECB29-8 (28K33) ECB29-8CB (28K34) 4 lbs. (2 kg)	2 steps	208	6.0	20,500	3.8	41	- - - -	- - - -
			220	6.7	22,900	3.8	43	- - - -	- - - -
			230	7.3	25,100	3.8	44	- - - -	- - - -
			240	8.0	27,300	3.8	47	- - - -	- - - -
	10 kW — ECB29-10 (28K35) ECB29-10CB (28K36) 5 lbs. (2 kg)	2 steps	208	7.5	25,600	3.8	50	- - - -	- - - -
			220	8.4	28,700	3.8	53	- - - -	- - - -
			230	9.2	31,400	3.8	55	- - - -	- - - -
			240	10.0	34,100	3.8	57	- - - -	- - - -
	12.5 kW — ECB29-12.5CB (28K37) 10 lbs. (5 kg)	3 steps	208	9.4	32,000	3.8	24	38	- - - -
			220	10.5	35,800	3.8	25	40	- - - -
			230	11.5	39,200	3.8	26	42	- - - -
			240	12.5	42,600	3.8	27	44	- - - -
	15 kW — ECB29-15CB (28K38) 10 lbs. (5 kg)	3 steps	208	11.3	38,400	3.8	28	45	- - - -
			220	12.6	43,000	3.8	29	48	- - - -
			230	13.5	47,000	3.8	30	49	- - - -
			240	15.0	51,200	3.8	31	52	- - - -
	20 kW — ECB29-20CB (28K39) 14 lbs. (6 kg)	4 steps	208	15.0	51,200	3.8	50	45	- - - -
			220	16.8	57,300	3.8	53	48	- - - -
			230	18.4	62,700	3.8	55	49	- - - -
			240	20.0	68,200	3.8	57	52	- - - -
	25 kW — ECB29-25CB (28K40) 18 lbs. (8 kg)	6 steps	208	18.8	64,100	3.8	42	38	38
			220	21.0	71,700	3.8	45	40	40
			230	23.0	78,300	3.8	46	42	42
			240	25.0	85,300	3.8	48	44	44
CB29M-65	5 kW — ECB29-5 (28K31) ECB29-5CB (28K32) 4 lbs. (2 kg)	1 step	208	3.8	12,800	4.6	29	- - - -	- - - -
			220	4.2	14,300	4.6	30	- - - -	- - - -
			230	4.6	15,700	4.6	31	- - - -	- - - -
			240	5.0	17,100	4.6	32	- - - -	- - - -
	8 kW — ECB29-8 (28K33) ECB29-8CB (28K34) 4 lbs. (2 kg)	2 steps	208	6.0	20,500	4.6	42	- - - -	- - - -
			220	6.7	22,900	4.6	44	- - - -	- - - -
			230	7.3	25,100	4.6	45	- - - -	- - - -
			240	8.0	27,300	4.6	48	- - - -	- - - -
	10 kW — ECB29-10 (28K35) ECB29-10CB (28K36) 5 lbs. (2 kg)	2 steps	208	7.5	25,600	4.6	51	- - - -	- - - -
			220	8.4	28,700	4.6	54	- - - -	- - - -
			230	9.2	31,400	4.6	56	- - - -	- - - -
			240	10.0	34,100	4.6	58	- - - -	- - - -
	12.5 kW — ECB29-12.5CB (28K37) 10 lbs. (5 kg)	3 steps	208	9.4	32,000	4.6	25	38	- - - -
			220	10.5	35,800	4.6	26	40	- - - -
			230	11.5	39,200	4.6	27	42	- - - -
			240	12.5	42,600	4.6	28	44	- - - -
	15 kW — ECB29-15CB (28K38) 10 lbs. (5 kg)	3 steps	208	11.3	38,400	4.6	28	45	- - - -
			220	12.6	43,000	4.6	30	48	- - - -
			230	13.5	47,000	4.6	30	49	- - - -
			240	15.0	51,200	4.6	32	52	- - - -
	20 kW — ECB29-20CB (28K39) 14 lbs. (6 kg)	4 steps	208	15.0	51,200	4.6	51	45	- - - -
			220	16.8	57,300	4.6	54	48	- - - -
			230	18.4	62,700	4.6	56	49	- - - -
			240	20.0	68,200	4.6	58	52	- - - -
	25 kW — ECB29-25CB (28K40) 18 lbs. (8 kg)	6 steps	208	18.8	64,100	4.6	43	38	38
			220	21.0	71,700	4.6	46	40	40
			230	23.0	78,300	4.6	47	42	42
			240	25.0	85,300	4.6	50	44	44
	30 kW — ECB29-30CB (28K41) 19 lbs. (9 kg)	6 steps	208	22.5	76,900	4.6	51	45	45
			220	25.2	86,000	4.6	54	48	48
			230	27.5	94,000	4.6	56	49	49
			240	30.0	102,400	4.6	58	52	52

*Refer to National or Canadian Electrical Code manual to determine wire, fuse and disconnect size requirements. Use wires suitable for at least 167°F (75°C).
**Electric heater capacity only — does not include additional blower motor heat capacity.
●Minimum circuit ampacity for blower motor only.

FIGURE 20.15

Electric heat data (1 phase). (*Courtesy of Lennox Industries, Inc.*)

Blower Coil Model Number	Electric Heat kW, Model Number & Shipping Weight		Number of Steps	Volts Input	kW Input	**Btuh Input	●Blower Motor Amps	*Minimum Circuit Ampacity	
								Circuit 1	Circuit 2
◆ 208/230v–3 ph ◆									
CB29M-41	8 kW	ECB29-8 (28K42) 5 lbs. (2 kg)	3 steps	208	6.0	20,500	2.4	24	– – – –
				220	6.7	22,900	2.4	25	– – – –
				230	7.3	25,100	2.4	26	– – – –
				240	8.0	27,300	2.4	27	– – – –
	10 kW	ECB29-10 (28K43) 6 lbs. (3 kg)	3 steps	208	7.5	25,600	2.4	29	– – – –
				220	8.4	28,700	2.4	31	– – – –
				230	9.2	31,400	2.4	32	– – – –
				240	10.0	34,100	2.4	33	– – – –
	15 kW	ECB29-15CB (28K44) 9 lbs. (4 kg)	3 steps	208	11.3	38,400	2.4	42	– – – –
				220	12.6	43,000	2.4	44	– – – –
				230	13.5	47,000	2.4	45	– – – –
				240	15.0	51,200	2.4	48	– – – –
CB29M-46	8 kW	ECB29-8 (28K42) 5 lbs. (2 kg)	3 steps	208	6.0	20,500	3.6	25	– – – –
				220	6.7	22,900	3.6	27	– – – –
				230	7.3	25,100	3.6	27	– – – –
				240	8.0	27,300	3.6	29	– – – –
	10 kW	ECB29-10 (28K43) 6 lbs. (3 kg)	3 steps	208	7.5	25,600	3.6	31	– – – –
				220	8.4	28,700	3.6	32	– – – –
				230	9.2	31,400	3.6	33	– – – –
				240	10.0	34,100	3.6	35	– – – –
	15 kW	ECB29-15CB (28K44) 9 lbs. (4 kg)	3 steps	208	11.3	38,400	3.6	44	– – – –
				220	12.6	43,000	3.6	46	– – – –
				230	13.5	47,000	3.6	47	– – – –
				240	15.0	51,200	3.6	50	– – – –
CB29M-51	8 kW	ECB29-8 (28K42) 5 lbs. (2 kg)	3 steps	208	6.0	20,500	3.8	26	– – – –
				220	6.7	22,900	3.8	27	– – – –
				230	7.3	25,100	3.8	28	– – – –
				240	8.0	27,300	3.8	29	– – – –
	10 kW	ECB29-10 (28K43) 6 lbs. (3 kg)	3 steps	208	7.5	25,600	3.8	31	– – – –
				220	8.4	28,700	3.8	32	– – – –
				230	9.2	31,400	3.8	34	– – – –
				240	10.0	34,100	3.8	35	– – – –
	15 kW	ECB29-15CB (28K44) 9 lbs. (4 kg)	3 steps	208	11.3	38,400	3.8	44	– – – –
				220	12.6	43,000	3.8	46	– – – –
				230	13.5	47,000	3.8	47	– – – –
				240	15.0	51,200	3.8	50	– – – –
	20 kW	ECB29-20CB (28K45) 19 lbs. (9 kg)	6 steps	208	15.0	51,200	3.8	31	26
				220	16.8	57,300	3.8	32	28
				230	18.4	62,700	3.8	34	29
				240	20.0	68,200	3.8	35	30
	25 kW	ECB29-25CB (28K46) 19 lbs. (9 kg)	6 steps	208	18.8	64,100	3.8	37	33
				220	21.0	71,700	3.8	39	34
				230	23.0	78,300	3.8	41	36
				240	25.0	85,300	3.8	43	38
CB29M-65	8 kW	ECB29-8 (28K42) 5 lbs. (2 kg)	3 steps	208	6.0	20,500	4.6	27	– – – –
				220	6.7	22,900	4.6	28	– – – –
				230	7.3	25,100	4.6	29	– – – –
				240	8.0	27,300	4.6	30	– – – –
	10 kW	ECB29-10 (28K43) 6 lbs. (3 kg)	3 steps	208	7.5	25,600	4.6	32	– – – –
				220	8.4	28,700	4.6	33	– – – –
				230	9.2	31,400	4.6	35	– – – –
				240	10.0	34,100	4.6	36	– – – –
	15 kW	ECB29-15CB (28K44) 9 lbs. (4 kg)	3 steps	208	11.3	38,400	4.6	45	– – – –
				220	12.6	43,000	4.6	47	– – – –
				230	13.5	47,000	4.6	48	– – – –
				240	15.0	51,200	4.6	51	– – – –
	20 kW	ECB29-20CB (28K45) 19 lbs. (9 kg)	6 steps	208	15.0	51,200	4.6	32	26
				220	16.8	57,300	4.6	33	28
				230	18.4	62,700	4.6	35	29
				240	20.0	68,200	4.6	36	30
	25 kW	ECB29-25CB (28K46) 19 lbs. (9 kg)	6 steps	208	18.8	64,100	4.6	38	33
				220	21.0	71,700	4.6	40	34
				230	23.0	78,300	4.6	42	36
				240	25.0	85,300	4.6	44	38

*Refer to National or Canadian Electrical Code manual to determine wire, fuse and disconnect size requirements. Use wires suitable for at least 167°F (75°C).
**Electric heater capacity only — does not include additional blower motor heat capacity.
●Minimum circuit ampacity for blower motor only.

FIGURE 20.16

Electric heat data (3 phase). (*Courtesy of Lennox Industries, Inc.*)

Blower Coil Model Number	Electric Heat kW, Model Number & Shipping Weight		Number of Steps	Volts Input	kW Input	**Btuh Input	●Blower Motor Amps	*Minimum Circuit Ampacity
				◆ 460v–3 ph ◆				
CB29M-41	10 kW	ECB29-10 (28K47) 12 lbs. (5 kg)	3 steps	440	8.4	28,700	1.3	15
				460	9.2	31,400	1.3	16
				480	10.0	34,100	1.3	17
	15 kW	ECB29-15 (28K48) 12 lbs. (5 kg)	3 steps	440	12.6	43,000	1.3	22
				460	13.8	47,000	1.3	23
				480	15.0	51,200	1.3	24
CB29M-51	10 kW	ECB29-10 (28K47) 12 lbs. (5 kg)	3 steps	440	8.4	28,700	1.9	16
				460	9.2	31,400	1.9	17
				480	10.0	34,100	1.9	17
	15 kW	ECB29-15 (28K48) 12 lbs. (5 kg)	3 steps	440	12.6	43,000	1.9	23
				460	13.8	47,000	1.9	24
				480	15.0	51,200	1.9	25
	20 kW	ECB29-20 (28K49) 18 lbs. (8 kg)	3 steps	440	16.8	57,300	1.9	30
				460	18.4	62,700	1.9	31
				480	20.0	68,200	1.9	32
	25 kW	ECB29-25 (28K50) 18 lbs. (8 kg)	3 steps	440	21.0	71,700	1.9	37
				460	23.0	78,300	1.9	39
				480	25.0	85,300	1.9	40
CB29M-65	10 kW	ECB29-10 (28K47) 12 lbs. (5 kg)	3 steps	440	8.4	28,700	2.3	17
				460	9.2	31,400	2.3	17
				480	10.0	34,100	2.3	18
	15 kW	ECB29-15 (28K48) 12 lbs. (5 kg)	3 steps	440	12.6	43,000	2.3	24
				460	13.8	47,000	2.3	25
				480	15.0	51,200	2.3	25
	20 kW	ECB29-20 (28K49) 18 lbs. (8 kg)	3 steps	440	16.8	57,300	2.3	30
				460	18.4	62,700	2.3	32
				480	20.0	68,200	2.3	33
	25 kW	ECB29-25 (28K50) 18 lbs. (8 kg)	3 steps	440	21.0	71,700	2.3	37
				460	23.0	78,300	2.3	39
				480	25.0	85,300	2.3	40
				◆ †575v–3 ph ◆				
†CB29M-51	20 kW	ECB29-20 (28K51) 18 lbs. (8 kg)	3 steps	550	16.8	57,300	††1.9	24
				575	18.4	62,700	††1.9	26
				600	20.0	68,200	††1.9	26
	25 kW	ECB29-25 (28K52) 18 lbs. (8 kg)	3 steps	550	21.0	71,700	††1.9	30
				575	23.0	78,300	††1.9	31
				600	25.0	85,300	††1.9	32
†CB29M-65	20 kW	ECB29-20 (28K51) 18 lbs. (8 kg)	3 steps	550	16.8	57,300	††2.3	25
				575	18.4	62,700	††2.3	26
				600	20.0	68,200	††2.3	27
	25 kW	ECB29-25 (28K52) 18 lbs. (8 kg)	3 steps	550	21.0	71,700	††2.3	30
				575	23.0	78,300	††2.3	32
				600	25.0	85,300	††2.3	33

*Refer to National or Canadian Electrical Code manual to determine wire, fuse and disconnect size requirements. Use wires suitable for at least 167°F (75°C).
**Electric heater capacity only — does not include additional blower motor heat capacity.
●Minimum circuit ampacity for blower motor only.
†NOTE – ALL 575v ELECTRIC HEATERS ARE USED WITH CB29M-51 & C829M-65 460v MODEL BLOWER COIL UNITS – A STEP-DOWN TRANSFORMER FOR THE BLOWER COIL UNIT IS FURNISHED WITH ALL 575v ELECTRIC HEATERS.
††Blower motor is rated at 460v.

FIGURE 20.17

Electric heat data (3 phase). (*Courtesy of Lennox Industries, Inc.*)

Blower Coil Model Number	Electric Heat kW, Model Number & Shipping Weight		Number of Steps	Volts Input	kW Input	[1]Btuh Input	[2]Blower Motor Amps	†Minimum Circuit Ampacity	
								Circuit 1	Circuit 2
◆ 208/230v–1 ph ◆									
CB30M-21/26 CB30U-21/26	2.5 kW	ECB29-2.5 (28K30) 4 lbs. (2 kg)	1 step	208	1.9	6,400	1.5	13	----
				220	2.1	7,200	1.5	14	----
				230	2.3	7,800	1.5	14	----
				240	2.5	8,500	1.5	15	----
	5 kW	ECB29-5 (28K31) ECB29-5CB (28K32) 4 lbs. (2 kg)	1 step	208	3.8	12,800	1.5	25	----
				220	4.2	14,300	1.5	26	----
				230	4.6	15,700	1.5	27	----
				240	5.0	17,100	1.5	28	----
	8 kW	ECB29-8 (28K33) ECB29-8CB (28K34) 5 lbs. (2 kg)	2 steps	208	6.0	20,500	1.5	38	----
				220	6.7	22,900	1.5	40	----
				230	7.3	25,100	1.5	42	----
				240	8.0	27,300	1.5	44	----
	10 kW	ECB29-10 (28K35) ECB29-10CB (28K36) 5 lbs. (2 kg)	2 steps	208	7.5	25,600	1.5	47	----
				220	8.4	28,700	1.5	50	----
				230	9.2	31,400	1.5	52	----
				240	10.0	34,100	1.5	54	----
CB30M-31 CB30U-31	5 kW	ECB29-5 (28K31) ECB29-5CB (28K32) 4 lbs. (2 kg)	1 step	208	3.8	12,800	1.73	25	----
				220	4.2	14,300	1.73	26	----
				230	4.6	15,700	1.73	27	----
				240	5.0	17,100	1.73	28	----
	8 kW	ECB29-8 (28K33) ECB29-8CB (28K34) 5 lbs. (2 kg)	2 steps	208	6.0	20,500	1.73	38	----
				220	6.7	22,900	1.73	40	----
				230	7.3	25,100	1.73	42	----
				240	8.0	27,300	1.73	44	----
	10 kW	ECB29-10 (28K35) ECB29-10CB (28K36) 5 lbs. (2 kg)	2 steps	208	7.5	25,600	1.73	47	----
				220	8.4	28,700	1.73	50	----
				230	9.2	31,400	1.73	52	----
				240	10.0	34,100	1.73	54	----
	12.5 kW	ECB29-12.5CB (28K37) 10 lbs. (5 kg)	3 steps	208	9.4	32,000	1.73	21	38
				220	10.5	35,800	1.73	22	40
				230	11.5	39,200	1.73	23	42
				240	12.5	42,600	1.73	24	43
	15 kW	ECB29-15CB (28K38) 10 lbs. (5 kg)	3 steps	208	11.3	38,400	1.73	25	45
				220	12.6	43,000	1.73	26	48
				230	13.5	47,000	1.73	27	52
				240	15.0	51,200	1.73	28	52

†Refer to National or Canadian Electrical Code manual to determine wire, fuse and disconnect size requirements. Use wires suitable for at least 167°F (75°C).
[1]Electric heater capacity only — does not include additional blower motor heat capacity.
[2]Minimum circuit ampacity for blower motor only.

FIGURE 20.18

Electric heat data (1 phase). (*Courtesy of Lennox Industries, Inc.*)

Blower Coil Model Number	Electric Heat kW, Model Number & Shipping Weight	Number of Steps	Volts Input	kW Input	[1]Btuh Input	[2]Blower Motor Amps	†Minimum Circuit Ampacity	
							Circuit 1	Circuit 2
			◆ 208/230v--1 ph ◆					
CB30M-41	5 kW / ECB29-5 (28K31) ECB29-5CB (28K32) 4 lbs. (2 kg)	1 step	208	3.8	12,800	1.73	25	----
			220	4.2	14,300	1.73	26	----
			230	4.6	15,700	1.73	27	----
			240	5.0	17,100	1.73	28	----
	8 kW / ECB29-8 (28K33) ECB29-8CB (28K34) 5 lbs. (2 kg)	2 steps	208	6.0	20,500	1.73	38	----
			220	6.7	22,900	1.73	40	----
			230	7.3	25,100	1.73	42	----
			240	8.0	27,300	1.73	44	----
	10 kW / ECB29-10 (28K35) ECB29-10CB (28K36) 5 lbs. (2 kg)	2 steps	208	7.5	25,600	1.73	47	----
			220	8.4	28,700	1.73	50	----
			230	9.2	31,400	1.73	52	----
			240	10.0	34,100	1.73	54	----
	12.5 kW / ECB29-12.5CB (28K37) 10 lbs. (5 kg)	3 steps	208	9.4	32,000	1.73	21	38
			220	10.5	35,800	1.73	22	40
			230	11.5	39,200	1.73	23	42
			240	12.5	42,600	1.73	24	43
	15 kW / ECB29-15CB (28K38) 10 lbs. (5 kg)	3 steps	208	11.3	38,400	1.73	25	45
			220	12.6	43,000	1.73	26	48
			230	13.5	47,000	1.73	27	50
			240	15.0	51,200	1.73	28	52
	20 kW / ECB29-20CB (28K39) 14 lbs. (6 kg)	4 steps	208	15.0	51,200	1.73	47	45
			220	16.8	57,300	1.73	50	48
			230	18.4	62,700	1.73	52	50
			240	20.0	68,200	1.73	54	52
CB30M-46 CB30U-41/46	5 kW / ECB29-5 (28K31) ECB29-5CB (28K32) 4 lbs. (2 kg)	1 step	208	3.8	12,800	2.4	26	----
			220	4.2	14,300	2.4	27	----
			230	4.6	15,700	2.4	28	----
			240	5.0	17,100	2.4	29	----
	8 kW / ECB29-8 (28K33) ECB29-8CB (28K34) 5 lbs. (2 kg)	2 steps	208	6.0	20,500	2.4	39	----
			220	6.7	22,900	2.4	41	----
			230	7.3	25,100	2.4	43	----
			240	8.0	27,300	2.4	45	----
	10 kW / ECB29-10 (28K35) ECB29-10CB (28K36) 5 lbs. (2 kg)	2 steps	208	7.5	25,600	2.4	48	----
			220	8.4	28,700	2.4	51	----
			230	9.2	31,400	2.4	53	----
			240	10.0	34,100	2.4	55	----
	12.5 kW / ECB29-12.5CB (28K38) 10 lbs. (5 kg)	3 steps	208	9.4	32,000	2.4	22	38
			220	10.5	35,800	2.4	23	40
			230	11.5	39,200	2.4	24	42
			240	12.5	42,600	2.4	25	43
	15 kW / ECB29-15CB (28K38) 10 lbs. (5 kg)	3 steps	208	11.3	38,400	2.4	26	45
			220	12.6	43,000	2.4	27	48
			230	13.5	47,000	2.4	28	50
			240	15.0	51,200	2.4	29	52
	20 kW / ECB29-20CB (28K39) 14 lbs. (6 kg)	4 steps	208	15.0	51,200	2.4	48	45
			220	16.8	57,300	2.4	51	48
			230	18.4	62,700	2.4	53	50
			240	20.0	68,200	2.4	55	52

†Refer to National or Canadian Electrical Code manual to determine wire, fuse and disconnect size requirements. Use wires suitable for at least 167°F (75°C).
[1]Electric heater capacity only — does not include additional blower motor heat capacity. [2]Minimum circuit ampacity for blower motor only.

FIGURE 20.19

Electric heat data (1 phase). (*Courtesy of Lennox Industries, Inc.*)

Blower Coil Model Number	Electric Heat kW, Model Number & Shipping Weight	Number of Steps	Volts Input	kW Input	[1]Btuh Input	[2]Blower Motor Amps	†Minimum Circuit Ampacity		
							Circuit 1	Circuit 2	Circuit 3
				◆ 208/230v–1 ph ◆					
CB30M-51 CB30U-51	5 kW	ECB29-5 (28K31) ECB29-5CB (28K32) 4 lbs. (2 kg)	1 step	208	3.8	12,800	2.4	26	----	----
				220	4.2	14,300	2.4	27	----	----
				230	4.6	15,700	2.4	28	----	----
				240	5.0	17,100	2.4	29	----	----
	8 kW	ECB29-8 (28K33) ECB29-8CB (28K34) 4 lbs. (2 kg)	2 steps	208	6.0	20,500	2.4	39	----	----
				220	6.7	22,900	2.4	41	----	----
				230	7.3	25,100	2.4	43	----	----
				240	8.0	27,300	2.4	45	----	----
	10 kW	ECB29-10 (28K35) ECB29-10CB (28K36) 5 lbs. (2 kg)	2 steps	208	7.5	25,600	2.4	48	----	----
				220	8.4	28,700	2.4	51	----	----
				230	9.2	31,400	2.4	53	----	----
				240	10.0	34,100	2.4	55	----	----
	12.5 kW	ECB29-12.5CB (28K37) 10 lbs. (5 kg)	3 steps	208	9.4	32,000	2.4	22	38	----
				220	10.5	35,800	2.4	23	40	----
				230	11.5	39,200	2.4	24	42	----
				240	12.5	42,600	2.4	25	43	----
	15 kW	ECB29-15CB (28K38) 10 lbs. (5 kg)	3 steps	208	11.3	38,400	2.4	26	45	----
				220	12.6	43,000	2.4	27	48	----
				230	13.5	47,000	2.4	28	49	----
				240	15.0	51,200	2.4	29	52	----
	20 kW	ECB29-20CB (28K39) 14 lbs. (6 kg)	4 steps	208	15.0	51,200	2.4	48	45	----
				220	16.8	57,300	2.4	51	48	----
				230	18.4	62,700	2.4	53	49	----
				240	20.0	68,200	2.4	55	52	----
	25 kW	ECB29-25CB (28K40) 18 lbs. (8 kg)	5 steps	208	18.8	64,100	2.4	41	38	38
				220	21.0	71,700	2.4	43	40	40
				230	23.0	78,300	2.4	45	42	42
				240	25.0	85,300	2.4	47	43	43
CB30M-65 CB30U-65	5 kW	ECB29-5 (28K31) ECB29-5CB (28K32) 4 lbs. (2 kg)	1 step	208	3.8	12,800	3.9	28	----	----
				220	4.2	14,300	3.9	29	----	----
				230	4.6	15,700	3.9	30	----	----
				240	5.0	17,100	3.9	31	----	----
	8 kW	ECB29-8 (28K33) ECB29-8CB (28K34) 4 lbs. (2 kg)	2 steps	208	6.0	20,500	3.9	41	----	----
				220	6.7	22,900	3.9	43	----	----
				230	7.3	25,100	3.9	45	----	----
				240	8.0	27,300	3.9	47	----	----
	10 kW	ECB29-10 (28K35) ECB29-10CB (28K36) 5 lbs. (2 kg)	2 steps	208	7.5	25,600	3.9	50	----	----
				220	8.4	28,700	3.9	53	----	----
				230	9.2	31,400	3.9	55	----	----
				240	10.0	34,100	3.9	57	----	----
	12.5 kW	ECB29-12.5CB (28K37) 10 lbs. (5 kg)	3 steps	208	9.4	32,000	3.9	24	38	----
				220	10.5	35,800	3.9	25	40	----
				230	11.5	39,200	3.9	26	42	----
				240	12.5	42,600	3.9	27	43	----
	15 kW	ECB29-15CB (28K38) 10 lbs. (5 kg)	3 steps	208	11.3	38,400	3.9	28	45	----
				220	12.6	43,000	3.9	29	48	----
				230	13.5	47,000	3.9	29	50	----
				240	15.0	51,200	3.9	31	52	----
	20 kW	ECB29-20CB (28K39) 14 lbs. (6 kg)	4 steps	208	15.0	51,200	3.9	50	45	----
				220	16.8	57,300	3.9	53	48	----
				230	18.4	62,700	3.9	55	50	----
				240	20.0	68,200	3.9	57	52	----
	25 kW	ECB29-25CB (28K40) 18 lbs. (8 kg)	5 steps	208	18.8	64,100	3.9	43	38	38
				220	21.0	71,700	3.9	45	40	40
				230	23.0	78,300	3.9	47	42	42
				240	25.0	85,300	3.9	48	43	43
	30 kW	ECB29-30CB (28K41) 19 lbs. (9 kg)	5 steps	208	22.5	76,900	3.9	50	45	45
				220	25.2	86,000	3.9	53	48	48
				230	27.5	94,000	3.9	55	50	50
				240	30.0	102,400	3.9	57	52	52

†Refer to National or Canadian Electrical Code manual to determine wire, fuse and disconnect size requirements. Use wires suitable for at least 167°F (75°C).
[1] Electric heater capacity only — does not include additional blower motor heat capacity.
[2] Minimum circuit ampacity for blower motor only.

FIGURE 20.20

Electric heat data (1 phase). (*Courtesy of Lennox Industries, Inc.*)

Blower Coil Model Number	Electric Heat kW, Model Number & Shipping Weight		Number of Steps	Volts Input	kW Input	[1] Btuh Input	[2] Blower Motor Amps	†Minimum Circuit Ampacity	
								Circuit 1	Circuit 2
◆ 208/230v–3 ph ◆									
CB30M-41	8 kW	ECB29-8 (28K42) 5 lbs. (2 kg)	3 steps	208	6.0	20,500	1.73	23	----
				220	6.7	22,900	1.73	24	----
				230	7.3	25,100	1.73	25	----
				240	8.0	27,300	1.73	26	----
	10 kW	ECB29-10 (28K43) 6 lbs. (3 kg)	3 steps	208	7.5	25,600	1.73	28	----
				220	8.4	28,700	1.73	30	----
				230	9.2	31,400	1.73	31	----
				240	10.0	34,100	1.73	32	----
	15 kW	ECB29-15CB (28K44) 9 lbs. (4 kg)	3 steps	208	11.3	38,400	1.73	41	----
				220	12.6	43,000	1.73	44	----
				230	13.5	47,000	1.73	45	----
				240	15.0	51,200	1.73	47	----
CB30M-46 CB30U-41/46	8 kW	ECB29-8 (28K42) 5 lbs. (2 kg)	3 steps	208	6.0	20,500	2.4	24	----
				220	6.7	22,900	2.4	25	----
				230	7.3	25,100	2.4	26	----
				240	8.0	27,300	2.4	27	----
	10 kW	ECB29-10 (28K43) 6 lbs. (3 kg)	3 steps	208	7.5	25,600	2.4	29	----
				220	8.4	28,700	2.4	31	----
				230	9.2	31,400	2.4	32	----
				240	10.0	34,100	2.4	33	----
	15 kW	ECB29-15CB (28K44) 9 lbs. (4 kg)	3 steps	208	11.3	38,400	2.4	42	----
				220	12.6	43,000	2.4	44	----
				230	13.5	47,000	2.4	45	----
				240	15.0	51,200	2.4	48	----
CB30M-51 CB30U-51	8 kW	ECB29-8 (28K42) 5 lbs. (2 kg)	3 steps	208	6.0	20,500	2.4	24	----
				220	6.7	22,900	2.4	25	----
				230	7.3	25,100	2.4	26	----
				240	8.0	27,300	2.4	27	----
	10 kW	ECB29-10 (28K43) 6 lbs. (3 kg)	3 steps	208	7.5	25,600	2.4	29	----
				220	8.4	28,700	2.4	31	----
				230	9.2	31,400	2.4	32	----
				240	10.0	34,100	2.4	33	----
	15 kW	ECB29-15CB (28K44) 9 lbs. (4 kg)	3 steps	208	11.3	38,400	2.4	42	----
				220	12.6	43,000	2.4	44	----
				230	13.5	47,000	2.4	45	----
				240	15.0	51,200	2.4	48	----
	20 kW	ECB29-20CB (28K45) 19 lbs. (9 kg)	6 steps	208	15.0	51,200	2.4	29	26
				220	16.8	57,300	2.4	31	28
				230	18.4	62,700	2.4	32	29
				240	20.0	68,200	2.4	33	30
	25 kW	ECB29-25CB (28K46) 19 lbs. (9 kg)	6 steps	208	18.8	64,100	2.4	36	33
				220	21.0	71,700	2.4	37	34
				230	23.0	78,300	2.4	39	36
				240	25.0	85,300	2.4	41	38
CB30M-65 CB30U-65	8 kW	ECB29-8 (28K42) 5 lbs. (2 kg)	3 steps	208	6.0	20,500	3.9	26	----
				220	6.7	22,900	3.9	27	----
				230	7.3	25,100	3.9	28	----
				240	8.0	27,300	3.9	29	----
	10 kW	ECB29-10 (28K43) 6 lbs. (3 kg)	3 steps	208	7.5	25,600	3.9	31	----
				220	8.4	28,700	3.9	32	----
				230	9.2	31,400	3.9	34	----
				240	10.0	34,100	3.9	35	----
	15 kW	ECB29-15CB (28K44) 9 lbs. (4 kg)	3 steps	208	11.3	38,400	3.9	44	----
				220	12.6	43,000	3.9	46	----
				230	13.5	47,000	3.9	47	----
				240	15.0	51,200	3.9	50	----
	20 kW	ECB29-20CB (28K45) 19 lbs. (9 kg)	6 steps	208	15.0	51,200	3.9	31	26
				220	16.8	57,300	3.9	32	28
				230	18.4	62,700	3.9	34	29
				240	20.0	68,200	3.9	35	30
	25 kW	ECB29-25CB (28K46) 19 lbs. (9 kg)	6 steps	208	18.8	64,100	3.9	38	33
				220	21.0	71,700	3.9	39	34
				230	23.0	78,300	3.9	41	36
				240	25.0	85,300	3.9	43	38

†Refer to National or Canadian Electrical Code manual to determine wire, fuse and disconnect size requirements. Use wires suitable for at least 167°F (75°C).
[1] Electric heater capacity only — does not include additional blower motor heat capacity.
[2] Minimum circuit ampacity for blower motor only.

FIGURE 20.21

Electric heat data (3 phase). (*Courtesy of Lennox Industries, Inc.*)

Blower Coil Model Number	Electric Heat kW, Model Number & Shipping Weight	Number of Steps	Volts Input	kW Input	[1]Btuh Input	[2]Blower Motor Amps	†Minimum Circuit Ampacity
			← 460v–3 ph →				
CB30M-41	10 kW ECB29-10 (28K47) 12 lbs. (5 kg)	3 steps	440	8.4	28,700	1.1	15
			460	9.2	31,400	1.1	16
			480	10.0	34,100	1.1	16
	15 kW ECB29-15 (28K48) 12 lbs. (5 kg)	3 steps	440	12.6	43,000	1.1	22
			460	13.8	47,000	1.1	23
			480	15.0	51,200	1.1	24
CB30M-51	10 kW ECB29-10 (28K47) 12 lbs. (5 kg)	3 steps	440	8.4	28,700	1.3	16
			460	9.2	31,400	1.3	16
			480	10.0	34,100	1.3	17
	15 kW ECB29-15 (28K48) 12 lbs. (5 kg)	3 steps	440	12.6	43,000	1.3	22
			460	13.8	47,000	1.3	23
			480	15.0	51,200	1.3	24
	20 kW ECB29-20 (28K49) 18 lbs. (8 kg)	3 steps	440	16.8	57,300	1.3	29
			460	18.4	62,700	1.3	29
			480	20.0	68,200	1.3	32
	25 kW ECB29-25 (28K50) 18 lbs. (8 kg)	3 steps	440	21.0	71,700	1.3	36
			460	23.0	78,300	1.3	38
			480	25.0	85,300	1.3	39
CB30M-65	10 kW ECB29-10 (28K47) 12 lbs. (5 kg)	3 steps	440	8.4	28,700	1.3	16
			460	9.2	31,400	1.3	17
			480	10.0	34,100	1.3	17
	15 kW ECB29-15 (28K48) 12 lbs. (5 kg)	3 steps	440	12.6	43,000	1.9	23
			460	13.8	47,000	1.9	24
			480	15.0	51,200	1.9	25
	20 kW ECB29-20 (28K49) 18 lbs. (8 kg)	3 steps	440	16.8	57,300	1.9	30
			460	18.4	62,700	1.9	31
			480	20.0	68,200	1.9	32
	25 kW ECB29-25 (28K50) 18 lbs. (8 kg)	3 steps	440	21.0	71,700	1.9	37
			460	23.0	78,300	1.9	39
			480	25.0	85,300	1.9	40
			← [3]575v–3 ph →				
†CB30M-51	20 kW ECB29-20 (28K51) 18 lbs. (8 kg)	3 steps	550	16.8	57,300	[4]1.3	24
			575	18.4	62,700	[4]1.3	25
			600	20.0	68,200	[4]1.3	26
	25 kW ECB29-25 (28K52) 18 lbs. (8 kg)	3 steps	550	21.0	71,700	[4]1.3	29
			575	23.0	78,300	[4]1.3	31
			600	25.0	85,300	[4]1.3	32
†CB30M-65	20 kW ECB29-20 (28K51) 18 lbs. (8 kg)	3 steps	550	16.8	57,300	[4]1.9	24
			575	18.4	62,700	[4]1.9	26
			600	20.0	68,200	[4]1.9	26
	25 kW ECB29-25 (28K52) 18 lbs. (8 kg)	3 steps	550	21.0	71,700	[4]1.9	30
			575	23.0	78,300	[4]1.9	31
			600	25.0	85,300	[4]1.9	32

†Refer to National or Canadian Electrical Code manual to determine wire, fuse and disconnect size requirements. Use wires suitable for at least 167°F (75°C).
[1] Electric heater capacity only — does not include additional blower motor heat capacity.
[2] Minimum circuit ampacity for blower motor only.
[3] ALL 575v ELECTRIC HEATERS ARE USED WITH 460v CB30M-51 & CB30M-65 MODEL BLOWER COIL UNITS – A 575v to 460v STEP-DOWN TRANSFORMER FOR THE BLOWER COIL UNIT IS FURNISHED WITH ALL 575v ELECTRIC HEATERS.
[4] Blower motor is rated at 460v.

FIGURE 20.22

Electric heat data (3 phase). (*Courtesy of Lennox Industries, Inc.*)

Circuits for an Electric Forced Air Heating System

This chapter examines the electrical circuits for an electric forced air heating system. Even though this is an electric heating system, the circuits are very close to the electrical circuits found in many other forced air heating systems in this book.

Low-Voltage Circuit

The low-voltage circuit (24 V ac) is supplied by the 24-V ac step-down transformer. As you will remember for other chapters, this transformer converts the 110-V ac line voltage to 24 V ac to be used in the low-voltage circuits. This voltage is used by the thermostat circuit and the heat relay. In some units that operate on 460 to 575 V ac, this circuit also supplies low voltage to the door interlock circuit. This circuit is used to close the electrical circuit to the unit if the door is removed. This is used as a safety feature.

208/230-V Circuits

Most single-phase units use this voltage as the main source of power for the unit. This voltage is used to power the indoor blower and operate the electric heating element(s). It is important to remember that the blower should always be the first to come on and the last to go off. Figures 21.1 to 21.7 show these circuits for a typical 208/230-V unit.

460- and 575-V Circuits

On units that use 460 or 575 V ac, a series of step-down transformers are used to convert the higher voltage to lower voltage used in some of the circuits in these units. 575 V ac is routed to a step-down transformer that converts this line voltage to 460 V ac. This voltage is then used to power the indoor blower and to power another step-down transformer that converts the 460 V ac to 24 V ac to be used by the thermostat and electric heat.

On units that have a supply line voltage of 460 V ac, the only step-down transformer that is used converts this line voltage to 24 V ac for use by the thermostat and door interlock switch. Figures 21.8 and 21.9 show the 460- and 575-V ac wiring diagrams.

As you will see in the next chapter on the operation and maintenance of these electric forced air heating systems, they come in several different configurations from single phase to multiphase. However, the basic operation of these units is, for the most part, the same.

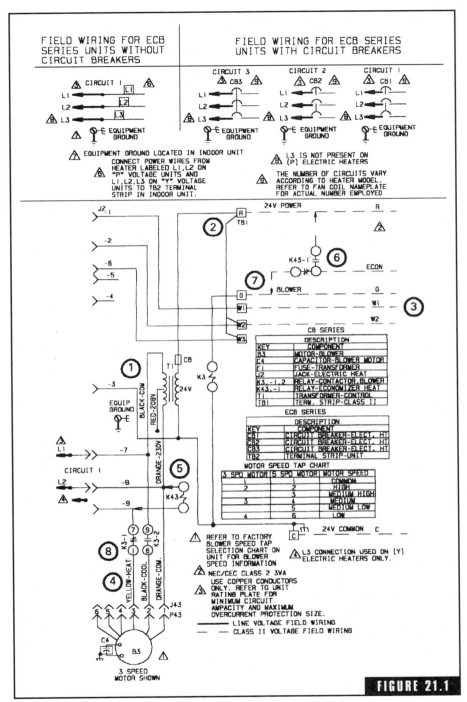

208/230-V ac sequence of operation. (*Courtesy of Lennox Industries, Inc.*)

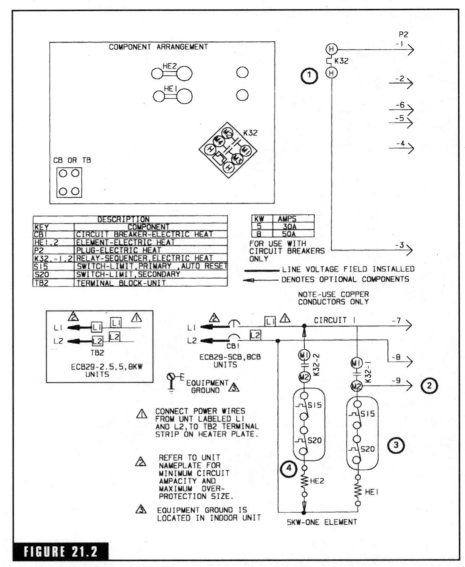

FIGURE 21.2

208/230-V ac single-phase wiring diagram. (*Courtesy of Lennox Industries, Inc.*)

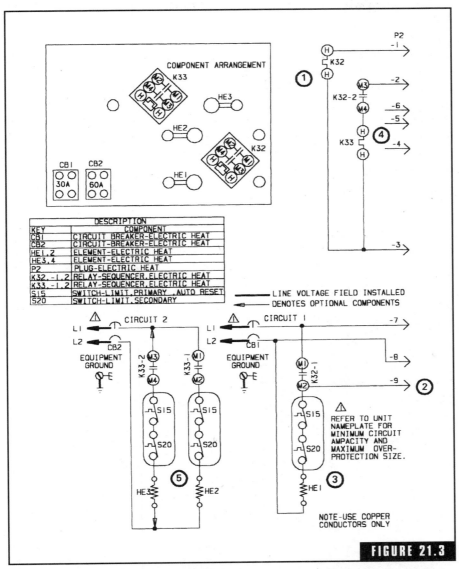

208/230-V ac wiring diagram with two-stage heat. (*Courtesy of Lennox Industries, Inc.*)

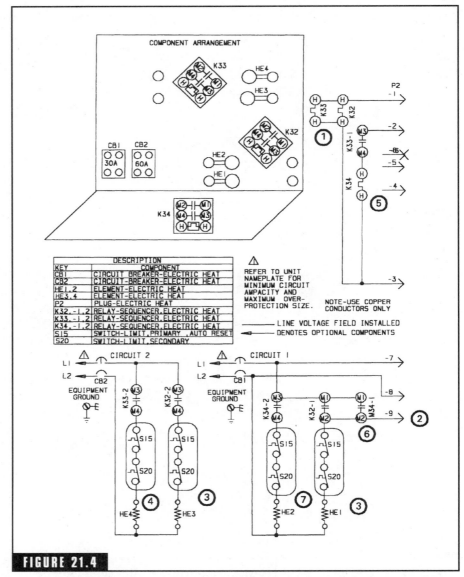

FIGURE 21.4

208/230-V ac single-phase two-stage operation. (*Courtesy of Lennox Industries, Inc.*)

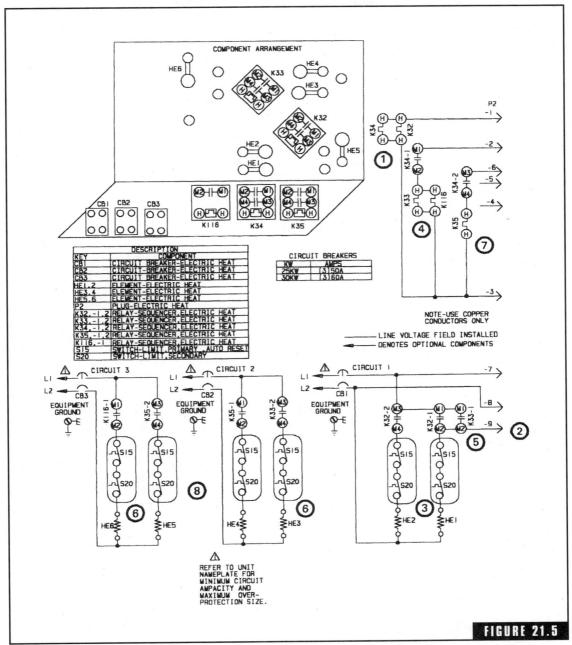

208/230-V ac single-phase three-stage operation. (*Courtesy of Lennox Industries, Inc.*)

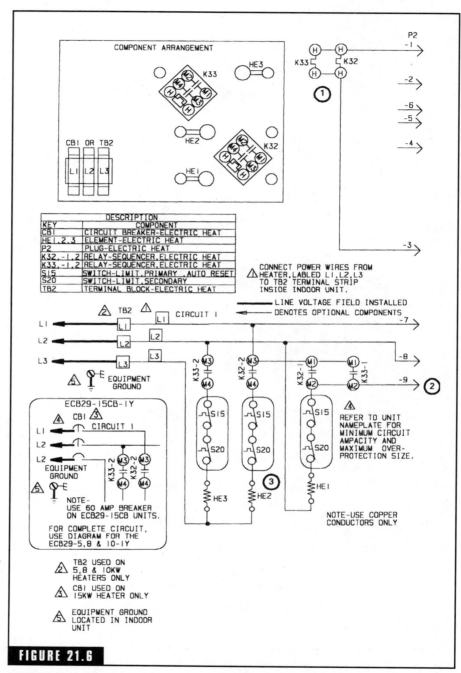

FIGURE 21.6

208/230-V ac three-phase single-stage operation. (*Courtesy of Lennox Industries, Inc.*)

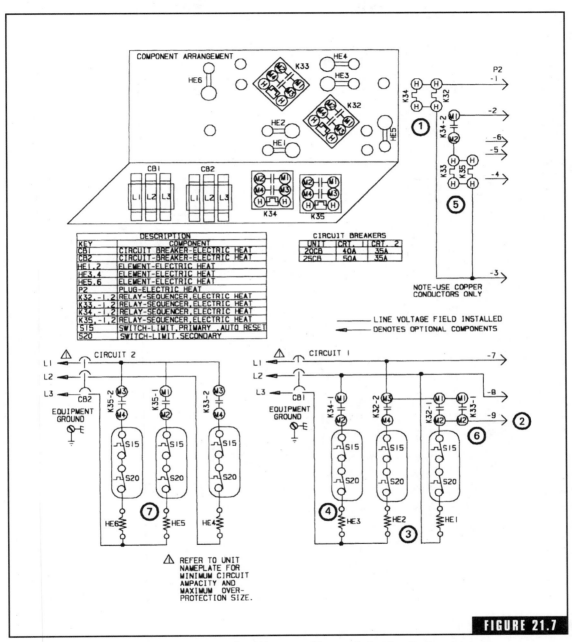

208/230-V ac three-phase two-stage operation. (*Courtesy of Lennox Industries, Inc.*)

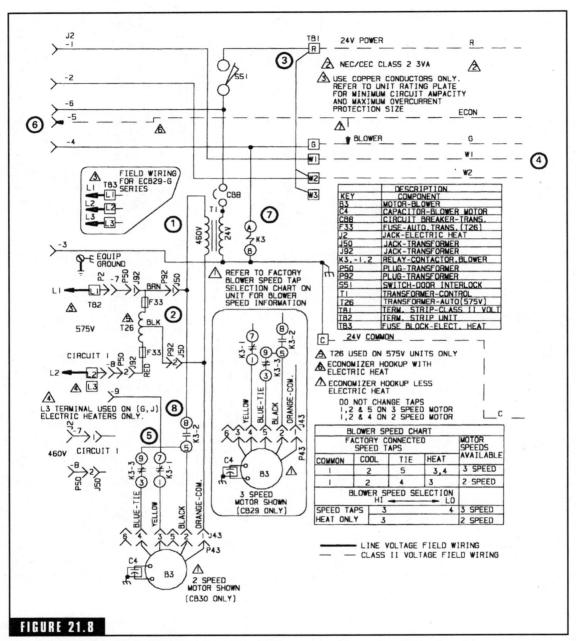

FIGURE 21.8

460- and 575-V ac three-phase single-stage operation. (*Courtesy of Lennox Industries, Inc.*)

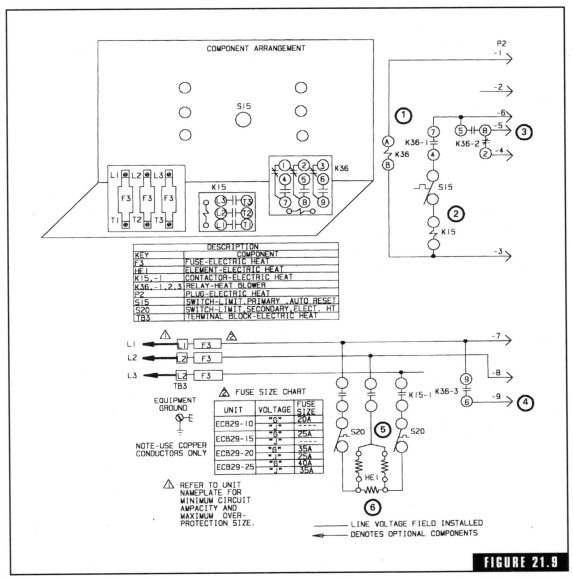

460- and 575-V ac three-phase single-stage operation. (*Courtesy of Lennox Industries, Inc.*)

Operation and Maintenance of Electric Forced Air Heating Systems

Electric forced air heating systems are simple to understand and maintain. Since the only moving parts in such a system are in the blower assembly, maintenance is fairly simple.

The operational sequence of an electric forced air heating system depends on whether the system is a single-phase or three-phase unit. We will examine the sequence of operation of each of these systems separately so that you will be familiar with how each works. Since you will be working with high line voltage, it is important to use caution when working on these units.

Single-Phase 208/230-V Sequence of Operation

As you will remember from the preceding chapter, there are several different types of single phase units. Some of the units you will be working on will have a single heating element, and some will have up to four or more elements. All these systems operate in the same basic way. The only difference is the sequence of the staging of when the elements are energized. On smaller units operating with one of two elements, the sequence of operation is listed below (see Fig. 21.1).

1. When the thermostat calls for heat, the electric heat relay K32 is energized with 24-V ac power.

2. When K32-1 closes, the blower is energized on heating speed.

3. If the primary limit switch (S15) and the secondary limit switch (S20) are both closed, heating element HE1 is energized, starting the heating cycle.

4. When K32-2 closes, assuming that the primary limit switch (S15) and the secondary limit switch (S20) are closed, the other electric element HE2 is energized.

5. Once the thermostat is satisfied, relays K32-1 and K32-2 stop the flow of electricity to the heating elements.

6. The blower will continue to run until the unit cools down to the set temperature.

Single-Phase 208/230-V Sequence of Operation with Second-Stage Heat [See Fig. 21.3]

1. When there is a call for heat, W1 of the thermostat energizes electric heat relay K32 with 24 V ac.

2. When K32-1 closes, the blower is energized on heating speed.

3. If the primary limit switch (S15) and secondary limit switch (S20) are closed, the heating element HE1 is energized to start the heating cycle.

4. In order for the second-stage heat to be active, you must remove the jumper between W2 and R on the thermostat.

5. When K32-2 closes, the unit is ready for the second-stage heat demand. When the demand for second-stage heat is sent by W2, electric heat relay K33 is energized with 24 V ac.

6. If the primary limit switch (S15) and the secondary limit switch (S20) are closed, electric heating elements HE2 and HE3 are energized.

7. Once the thermostat is satisfied, relay K32-2 opens, and the flow of electricity to heating elements HE2 and HE3 stops.

8. Heating element HE1 continues to operate until the signal is sent from W1 and the flow of electricity stops.

9. The blower continues to operate until the unit cools down.

Single-Phase 208/230-V Sequence of Operation with Three-Stage Heat (See Fig. 21.5)

1. When the thermostat calls for heat, W1 energizes the electric heat relays K32 and K34 with 24 V ac.

2. Relay K32-1 then closes, energizing the blower to operate on heating speed.

3. If the primary limit switch (S15) and the secondary limit switch (S20) are closed, K32-1 energizes heating element HE1, and K32-2 energizes heating element HE2.

4. In order for the second-stage heat to be active, you must remove the jumper between W2 and R on the thermostat.

5. When W2 calls for second-stage heat, relays K33 and K116 are energized with 24 V ac.

6. If the primary limit switch (S15) and the secondary limit switch (S20) are closed, relays K33-2 and K116-1 close, and power is sent to heating elements HE3 and HE6.

7. In order for third-stage heat to be active, you must remove the jumper between W3 and R on the thermostat.

8. When relay K34-2 closes, the unit is ready for third-stage heat demand. W3 sends a third-stage heat demand. This then energizes relay K35 with 24 V ac.

9. If the primary limit switch (S15) and the secondary limit switch (S20) are closed, electric heating elements HE4 and HE5 are energized.

10. As the thermostat becomes satisfied, the stages begin to deenergize, causing the different stages to shut down in reverse order. The blower will continue to run until the unit cools down.

The sequence of operation for the 460/575-V ac units is the same. The only difference is that the 575-V ac unit uses a step-down transformer to convert this high voltage to the lower voltage needed to operate the unit.

Maintaining an Electric Forced Air Heating System

Maintaining an electric forced air heating system is not as complex as some of the other types of forced air heating systems in this book. You do not have to concern yourselves with pilot lights, fire pots, oil tanks, etc. All that needs to be done to the electric forced air heating system are some simple maintenance functions.

Before attempting any maintenance on an electric forced air heating system, make sure that you turn off the power at the breaker panel or fuse box. The first item that needs attention is the filter. This should be changed several times during the heating season. The filter is located in the cold air plenum on the heating system before the blower. Be sure to look at the arrow on the filter so that you put the filter in properly. On most filters, there will be a wire cover over one end of the filter. This will be the side that faces the blower. This is done so that the filter material is not drawn into the blower. It is there for protection.

If the unit is a belt-drive unit, check the tension on the belt. You should have a deflection of ¼ to ½ in on the belt. If this is not the case, adjust the tension with the tension screw located on the motor mount frame until you get the proper tension.

If the blower motor has oil ports, place 1 or 2 drops of oil in these ports. They will be located on each end of the motor. If there are no oil holes on your motor, this is a sealed unit, and no oiling is necessary.

You should examine the blower cage for dirt and grease. If there is a buildup on the fins, they should be cleaned. If the fins get covered with dirt and grease, you will not get the proper flow of air, and the unit will not operate efficiently. Use a vacuum cleaner to clean these fins.

The other check that can be performed is to examine the system for any signs of frayed or bare wires. If you find any bad wires, not only must they be repaired, but the cause of the problem should be determined. The other items that should be checked, such as the amperage draw of the motor and limit switch operations, should be done very carefully because you are dealing with very high voltage (this definitely should *not* be attempted by the homeowner).

An item that can be done by the homeowner and will help to reduce the cost of operation is to seal the ductwork. You should take a look at the joints in the ductwork. If there are any joints that do not meet properly or if there are gaps in the seams, these should be sealed to keep heat from escaping. You can use gray tape to seal these joints. This simple maintenance item will return many times the cost of the repair.

I know that I have not gone into as much detail on this type of forced air heating unit as I did with the others, but this is not a very complex system to operate or maintain. If you change the filters, check the condition of the unit, and maintain the blower and motor, that is all that is needed to keep the unit in proper operating condition.

The next chapter covers troubleshooting such a system in the event of failure.

Troubleshooting Electric Forced Air Heating Systems

As you have seen from the preceding chapters on electric forced air heating systems, these systems have fewer parts than the other systems discussed in this book and require less maintenance. Troubleshooting these systems requires that you have a good understanding of the relationship between the parts and what can go wrong that will cause a system to fail. Since this is an electric forced air heating system and you will be working with high line voltage, care must be taken to ensure that the power is turned off before making repairs.

No Heat

As with most other forced air heating systems, this is the most common call that a service professional will receive. When dealing with an electric forced air heating system, power is everything. Find the fuse box or breaker panel. Since these units use a minimum of 208 V ac, in the breaker panel there will be a double breaker that is used to control the power. Check the panel to see if you can locate the breaker for the heating system. If none of the breakers or fuses is marked, you will need to start the blower manually to determine which fuse or breaker controls the system. To do this, you will have to turn the blower control switch on the thermostat to the "manual" setting. Figure 23.1 shows the location of this switch.

With the blower on manual, the blower should start. If the blower is running, go to the fuse box or breaker panel and turn off the breakers or remove the fuses until you locate the fuse or breaker that controls the unit. Mark this location on the panel for future reference.

If the blower did not start when you turned the switch to manual and you have checked to make sure that the breaker is turned on or the fuse is good, then you need to begin the troubleshooting process.

First, turn the power off to the unit. Next, return to the thermostat, and place the blower switch back to the "on" or "automatic" setting. Turn the thermostat up past the temperature in the room so that it is calling for heat. Remove the door to the heating section of the unit. You may need to "fool" the blower door interlock into thinking that the door is in place. Figure 23.2 shows the location of this switch. With the blower door removed, the unit will not operate in some cases, so you will need to hold the button in to check the operation.

Turn the power back on to the unit, and observe the operation. You are looking to see if the heating element(s) begin to heat. If there is no sign of the elements beginning to heat, you will have to check the primary and secondary limits to see if the unit has seen its high limit. If this is the case, there will be no activity with the heating elements. This is so because the primary limit switch is an autoreset type of limit, whereas the secondary high limit is a one-time limit switch and will have to be replaced if the unit has reached its high limit. Figure 23.3 shows the typical locations of the primary and secondary limits.

If you have determined that this is the problem, then you have found the cause of

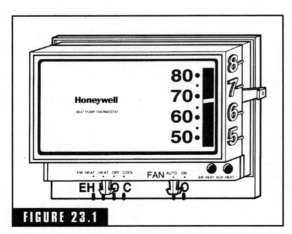

FIGURE 23.1

Heat-cool thermostat. (*Courtesy of Honeywell, Inc.*)

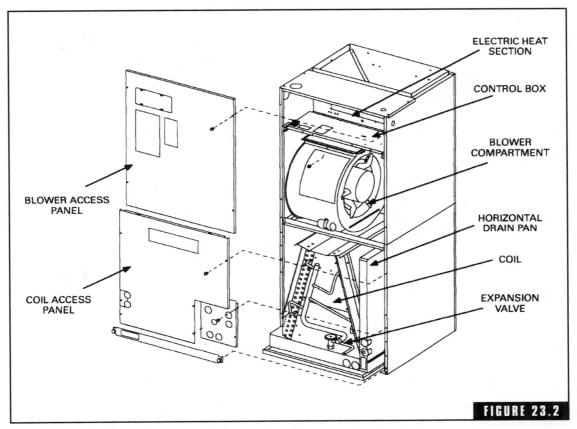

ELECTRIC HEAT
SECTION

CONTROL BOX

BLOWER
COMPARTMENT

HORIZONTAL
DRAIN PAN

COIL

EXPANSION
VALVE

BLOWER ACCESS
PANEL

COIL ACCESS
PANEL

FIGURE 23.2

Electric forced air heating system. (*Courtesy of Lennox Industries, Inc.*)

the no heat. As you will remember from preceding chapters, the sequence of operation is that the heating relays will sequence the blower and the heating elements to start. If the blower does not start, the unit will run until it reaches the high limit, and then it will shut off. This is a safety feature of the unit. This also will cause the secondary limit to fail, and it will have to be replaced before you can begin working on the blower.

To replace the secondary limit switch, first turn the power off to the unit. Next, you will have to determine what type of electric forced air heating system you are working on. As you will recall, there are several different configurations to the units depending on if it is a single-stage or multistage unit. On some single-stage units, there will be only

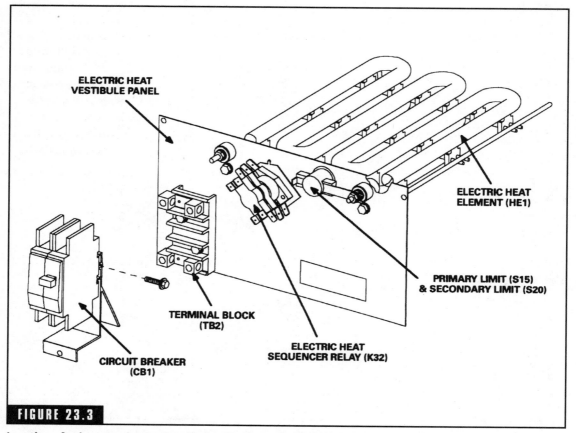

ELECTRIC HEAT
VESTIBULE PANEL

ELECTRIC HEAT
ELEMENT (HE1)

PRIMARY LIMIT (S15)
& SECONDARY LIMIT (S20)

TERMINAL BLOCK
(TB2)

ELECTRIC HEAT
SEQUENCER RELAY (K32)

CIRCUIT BREAKER
(CB1)

FIGURE 23.3

Location of primary and secondary limits. (*Courtesy of Lennox Industries, Inc.*)

one limit switch, whereas on other single-stage and multistage units, there will be two or more. You must determine which secondary limit has failed. You will need to perform a continuity check on these secondary limits to make this determination.

Remove the wires to the limit switch. Set your meter to the correct ohms setting, and place a probe on the terminals of the secondary limit, one on each side. If you receive an ohms reading, you have continuity, and the limit is working properly. If you do not get a reading, you have found the defective secondary limit. This will need to be replaced before the unit will operate. Remove the screws that hold the unit in place, and replace the secondary limit with one of the same voltage rating. Replace the wires, and make sure that all connections are tight.

Once you have the new secondary limit in place, check the operation of the unit again. If the heat sequencer relay still does not allow for the operation of the heating elements, turn the power off and check the other secondary limits, since you still have one that is defective. Once you have repaired this problem, check for the operation of the blower. If the blower starts, you have repaired the unit, and you should allow it to run for a complete cycle to check the operation.

If the blower does not start (as you will recall, the blower is always first on and last off), you will need to check the blower relay. Figure 23.4 shows the typical location of this relay. When the thermostat calls for heat, the relay is energized, allowing the blower to operate on heating speed. If the relay does not energize, the blower will not operate. Check the operation of the blower relay before moving on to the next step.

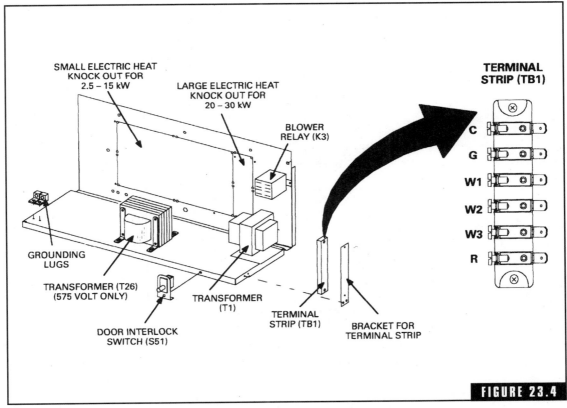

FIGURE 23.4

Location of the blower relay. (*Courtesy of Lennox Industries, Inc.*)

Check to make sure that you have power to the relay. If you do not have power to this relay, check to make sure that you have power to the unit. If there is no power, reset the breaker or replace the fuse. If there is still no power to the relay, check the thermostat to make sure that it is calling for heat. If the unit is calling for heat, you will need to check for power to the thermostat. Remove the thermostat from the subplate (Fig. 23.5). Check for power at W1 on the plate. If you have power at this location, replace the blower relay.

If you have checked for power at all the locations (i.e., relay, thermostat, and transformer) and the blower still does not operate, you will need to replace the blower motor.

Figure 23.6 shows the blower assembly. Turn the power off to the unit. Remove the bolts that hold the blower assembly to the unit. Remove the wires from the blower motor. (In some cases, the wires will be connected in a wire cabinet. You will have to disconnect them there in this case.) Remove the screw that holds the blower motor to the cage. Loosen the setscrew that holds the cage to the blower motor shaft. Slide the motor out of the blower housing. Remove the screws that hold the mounting bracket to the motor. Replace the motor with one of the same size and horsepower rating. Make sure that all electrical connections are tight after the motor is installed.

Once the motor is replaced, check the operation of the unit through a cycle. The other cause for a no heat could be the heating elements. After you have checked and/or replaced any defective parts and there is still no heat, the elements would be the next thing to check.

Since this is an electric resistance type of heater, you will need to check for continuity of the heating elements. Turn off the power to the unit. Remove any wires that are connected to the heating elements. Set your meter to ohms. Connect one end of the meter leads to one terminal of the heater and the other end to the other terminal. If you get no or a very low ohm reading, there is not enough resistance in the element, and it will have to be replaced.

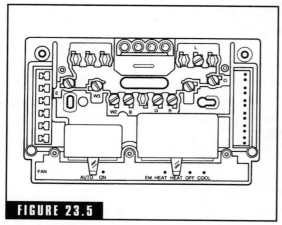

FIGURE 23.5

Thermostat backplate. (*Courtesy of Honeywell, Inc.*)

These elements will be located in a vestibule. You will have to remove the cover to this vestibule to gain access to the heating elements.

Once you have the vestibule removed, remove any limit switches and relays that are connected to the terminals. Remove the screws and/or nuts that hold the element to the vestibule. Remove the element, and replace is with one of the same size and wattage. You also must remember to check the voltage rating of the one that you are replacing and replace it with one of the same electrical rating.

Once you have the element replaced and the unit reassembled, check all connections before turning the power back on to the unit. Now turn the power to the unit back on, and check the operation. Run the unit through a cycle to make sure that the unit is operating properly.

Not Enough Heat

If you receive a call that the unit is not producing enough heat, you need to check for a couple of probable causes. The most common cause is a dirty air filter. Check the condition of the filter to make sure that it is clean. If the filter is dirty, replace it, and check the operation. You also will want to check the registers to make sure that they are open.

If the filter was not the cause, check to see if you are dealing with a multistage unit. If it is a multistage unit, you could have a problem with sequencing relays not operating properly to start the heating cycle on the balance of the heating elements.

Figure 23.7 shows a typical single-phase and three-phase multistage unit. Check the operation of the sequencing relays. As the unit begins the heating cycle, these relays energize to allow for the sequencing of the heating elements. If the relays are not operating properly, the additional heating elements will not come on, and the result will be that the unit is not putting out enough heat. The unit also will run longer that normal.

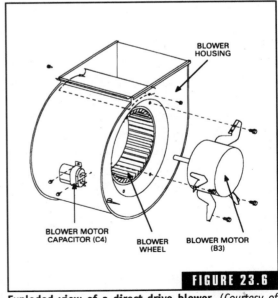

FIGURE 23.6

Exploded view of a direct-drive blower. (*Courtesy of Lennox Industries, Inc.*)

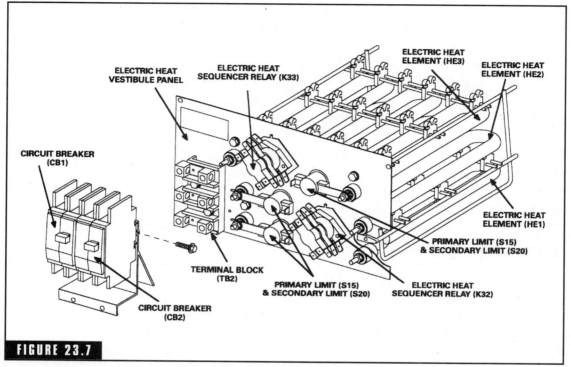

FIGURE 23.7

Single- and three-phase parts arrangement. (*Courtesy of Lennox Industries, Inc.*)

Check for power at the sequencing relay. If you have power but the elements are not operating, replace the relay. This assumes that the primary and secondary limits are operating properly. If no power is detected at the relay, you will need to check the condition of the primary and secondary limits. You will need to check for continuity at both these limits.

Turn the power off to the unit, and remove the wires from the primary and secondary limits. Set your meter to ohms, and check the continuity of the limits. If there is no continuity to a limit, replace it. Replace the wires, and turn the power back on to the unit. Check the operation of the unit. If you still do not have the other stage heating, check the thermostat. Since the thermostat is the device that determines when to begin the staging of the elements by sending the "signal" to begin the staging, you will want to check the condition of W2 on the subplate. If there is a problem with the subplate, replace it. Check the operation of the unit. It is always a good idea to check the wire diagram of the unit that you are working on to make sure that the sequence of operation is the same as described here and in the preceding chapters.

Conclusion

As we have discussed in this section, the electric forced air heating system is a much simpler device in some areas than the other forced air heating systems in this book. Let's take one last look at the sequence of operation before leaving this subject.

1. When there is a call for heat from the thermostat, this signal is sent to the electric heat relay.

2. When the relay closes, assuming that the primary and secondary limits are closed, the electric heat contactor closes.

3. The blower relay is then energized, and the blower starts.

4. Assuming that the primary and secondary limits are closed, the electric heating elements are energized, and the heating cycle begins.

5. When the thermostat is satisfied, the heating relay deenergizes, and the blower continues to run until the unit is cooled and the cycle ends.

If you remember this basic sequence of operation, it will help you to troubleshoot such a system more effectively.

Is a Heat Pump Right for You?

Is a heat pump the right choice for you? It is if you answer yes to the following question. Would you like a system that not only heats your home in the winter but also cools it in the summer? If the answer that you gave was yes, then a heat pump if right for you.

It does not matter what type of primary heating system you currently have in your home, a heat pump still will work. If you are currently using a heating system that does not require heat ducts, i.e., hot water, you will have to install an air mover and the needed ductwork to handle the airflow from the heat pump.

Let's take a moment to explore the different types of heating systems that are in use today and compare them with a heat pump.

Electric Forced Air Heat

The home that I currently own in Oregon is heated with electricity. This type of forced air heating produces heat very fast. It is not the warmest heat that is available, and it can be very expensive to operate.

Wood Heat

Wood heat is very warm and cozy in the winter. The smell of a good piece of wood burning is great. The drawback to wood heat is that you either have to go out and cut the wood, split it, and stack it, or you have to purchase it. This can be a great expense depending on the part of the country in which you live. I have had wood heat, and I do not enjoy having to haul the wood to the house. There is also the added expense of having the chimney cleaned. And you have to worry about the chance of a chimney fire. If the wood stove is on the main floor of the home, you also have the concern of someone getting burned on the hot stove.

Hot-Water Baseboard Heat

This is a great way to heat a home in the cold climate regions of the country. The heat is very warm, and you get the added benefit of moisture in the home. The drawback to this type of system is that it requires a great deal of space. You also have to worry about fuel deliveries if the unit is fueled by oil or propane. There is also no way to cool the home with this type of system.

Forced Air Heating Systems

This type of heating system is the most common type in the Northwest and other parts of the country. I love forced air heating. It has always served to keep me warm in the winter. This is also the best type of unit to have if you are considering (and I think you should) the addition of a heat pump.

All the heating systems listed here are good for heating your home, but they do nothing to help cool the home in the summer. It makes sense to consider a unit that not only will help to heat your home in the winter but also will cool it in the summer and have the added benefit of saving you money in the process.

Heat Pumps

There are several different types of heat pumps available. These include air source, ground source, and water source heat pumps.

Which type of unit to install is a decision that you will need to make depending on the area where you live. The most common type of heat pump is the air source heat pump. I will go into more detail on the different types of heat pumps in the next chapter.

When considering if a heat pump is right for you, just keep in mind that this is the only method available that can heat and cool your home with the same unit. It is also a great way to increase the value of your home at the time of resale.

How Does a Heat Pump Work?

As discussed in the preceding chapter, several types of heat pumps are available on the market. They all perform the same function, but they simply do this in different ways. Let's explore how a heat pump works.

The term *pump* implies that a heat pump moves heat from one point to another. In the winter, it moves heat from outside the home inside. In the summer it moves heat from inside the home outside. A heat pump works like a water pump. It pumps heat "up hill" to the home in the winter. Heat pumps use the refrigeration cycle to do this.

If you take a look at Figs. 25.1 and 25.2, I will explain how this refrigeration cycle works. A *refrigerant* is a fluid that vaporizes, or boils, at a very low temperature. This fluid moves through tubes that are called *refrigerant lines*. These lines travel throughout the heat pump. Figure 25.1 shows the refrigerant cycle used to heat a home.

We will begin at point *A*. At point *A* the refrigerant is a cold fluid—colder than the outside air. This fluid then flows to the outside coil (point *B*). This coil, or heat exchanger, has a very large surface area that will transfer the heat from the outside air to the fluid. The heat that is

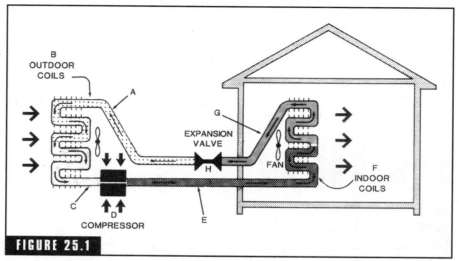

Heating cycle of an air source heat pump. (*Courtesy of the EWEB.*)

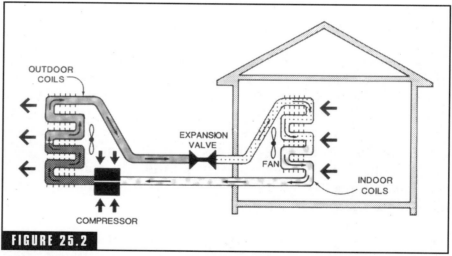

Cooling cycle of an air source heat pump. (*Courtesy of the EWEB.*)

absorbed by the fluid causes it to vaporize. The heat exchanger is also called an *evaporator coil* during this cycle. When a substance changes state from a liquid to a gas, a large amount of energy transfer takes place.

At point *C*, this vapor is now a very cold gas. This change was caused by the liquid being warmed by the outside air and changing to a gas. This cool gas could never warm the house in its current state, so a compressor is used. At point *D* the compressor raises the pressure of the gas. When the compressor squeezes the gas, the temperature of the gas begins to rise. The compressor is the heart of the heat pump. It is this device that performs most of the work. It is also the device that will force the now hot gas further "up hill" to point *E*.

Point *F,* the indoor coil, is where this now heated and compressed gas gives up its heat so that the air in the home can be heated. The blower fan blows air over the indoor coil to distribute the warm air into the home. As the gas begins to give off its heat, it also begins the cycle of changing back from a gas to a liquid. This is done in the indoor coil, which is also known as the *condenser coil.*

This mix of warm gas and cool liquid continues past point *G* to point *H.* At point *H,* the metering device reduces the pressure of the liquid to a point where it becomes cold again and is ready to absorb heat from the outside air to continue the heating cycle.

In Fig. 25.2, just the reverse is the case. Most heat pumps have a reversing valve that will reverse the flow of the liquid so that it can be used as a cooling device for the home. This is truly the advantage of a heat pump.

What I just described to you was the operation of an air source heat pump. Let's now look at the other two common types of heat pumps: ground source and water source.

Ground Source Heat Pumps

Ground source heat pumps, as the name implies, take their heat from the ground, where the temperature is more constant than the air. In some parts of the country where the outside air temperature is too low in the winter for an air source heat pump to operate efficiently, a ground source heat pump may be practical. Figure 25.3 shows a horizontal loop ground source heat pump.

A properly installed ground source heat pump may last much longer than an air source heat pump because there is a much smaller temperature difference because the ground maintains a more constant temperature throughout the year. This will cause the compressor to not have to work as hard, and ground source heat pumps do not have to defrost, which will save energy and improve operating efficiency.

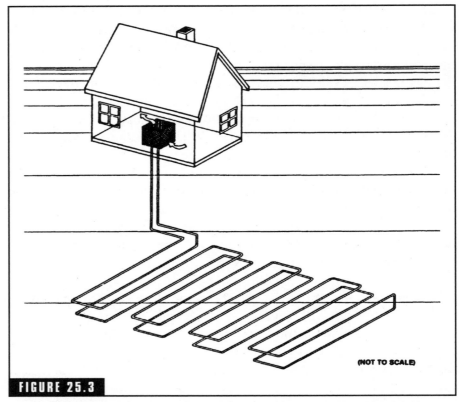

(NOT TO SCALE)

FIGURE 25.3

Multiple-layered horizontal ground source heat pump. (*Courtesy of the U.S. Department of Energy.*)

There are some disadvantages to a ground source heat pump as well. First is the initial cost. These units typically cost more to install than an air source heat pump. If you do not have a large enough heating load in the winter, you might not be able to offset the cost of installation of this type of unit.

A second disadvantage is that you may not be able to find someone in your area to install this type of unit. This also would cause you to have a long wait as well if you need service for the unit. The reason for this is that these types of units require some unique designs. Your site must be evaluated to make sure that the moisture content is correct. You must ensure that the ground around the home will not freeze in the winter when the heat pump extracts the heat from it.

A third disadvantage is that this type of installation requires excavation and drilling. There also may be some expensive landscaping work that will need to be done. Some of these units can just be drilled straight down, which will reduce the cost, but you must remember the cost to repair the pipe if this becomes necessary.

Water Source Heat Pumps

These types of heat pumps use some body of water as the heat source. Figure 25.4 shows a water source heat pump. Sources of heat for these units may be a well, lake, or stream. In any case, these units have the same advantage as ground source heat pumps in that the

(NOT TO SCALE)

FIGURE 25.4

Water source heat pump. (*Courtesy of the U.S. Department of Energy.*)

water will be warmer than the outside air temperature. These units also share one of the disadvantages with a ground source heat pump. They can be expensive to install. If you already have an existing well, this can help to keep the cost of installation down.

Before you decide to install a water source heat pump, you must have the area evaluated. If you are going to use a well as the heat source, you must make sure that you have the proper flow and that the water will be at the right temperature for the system to operate properly.

If you are going to use a lake or stream, you must first find out if it is legal to use this as a heat source. Some areas have environmental laws that prohibit the use of lakes and streams for heat. You also must make sure that the water does not get too cold to use.

The next problem is what to do with the water after the heat is extracted? In some areas you are required to dump the water into the sewer system. This will raise your bill. In other areas you are required to dump the water into another well with an aquifer. Make sure that you check your local laws before beginning this type of project.

As you have learned in this chapter, there are many choices when it comes to heat pumps. Each one has advantages and disadvantages. Which one is right for you depends on your personal needs and the part of the country in which you live. In any case, there are other things to consider before placing your money down for a heat pump. The next chapter introduces you to the terms used to determine the efficiency of the heat pumps and discuss the point at which a backup heating system is needed.

Introduction to Heat Pumps

Chapter 25 discussed heat pumps in general. There are many other factors involved in heat pumps that will be discussed in this chapter. Figure 26.1 shows the components of a heat pump.

COP

This term is known as *coefficient of performance* (COP). It is the rating given to a heat pump to tell the consumer the rate of heat output to electric input. Another way to look at this rating is to consider that if you are heating with electric resistance heat, forced air, or baseboard, you are getting a dollars worth of heat for every dollar that you spend. While this may sound like a great deal, consider that a heat pump with a COP rating of 2.5 is giving you $2.50 worth of heat for every $1.00 that you spend in electricity.

You must keep in mind that these numbers are based on testing under controlled conditions where the temperature indoors is 70°F and the outside temperature is 47°F. The best COP rating is 3. A number of factors will keep an air source heat pump from maintaining a COP rating of 3. One of these factors is the outdoor air temperature. Heat pumps that operate in winter will run most of the time at outdoor temperatures of less than 47°. As the temperature drops, the COP also

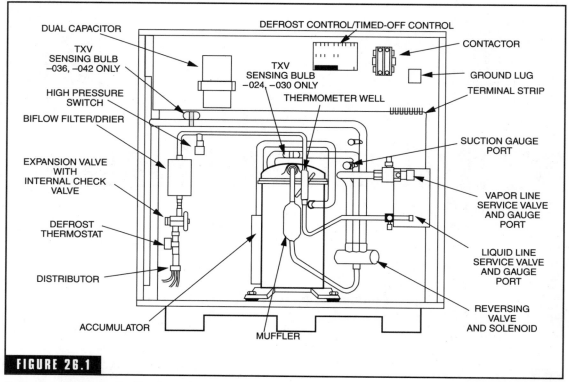

FIGURE 26.1

Heat pump components. (*Courtesy of Lennox Industries, Inc.*)

will drop. You can expect that if the temperature outdoors runs closer to 20°F, the COP number will be closer to 2 than 3.

EER

Energy efficiency rating (EER) is used to measure the cooling efficiency of a heat pump. Like the COP, the higher the EER rating, the better is the system. The more that you understand both the COP and the EER, the more informed you will be when the time comes to purchase a heat pump.

HSPF

The industry standard test for the overall heating efficiency is a rating called *heating season performance factor* (HSPF). This is a test conducted under laboratory conditions. It takes into account the reduction in energy caused by defrosting, temperature fluctuations, supplemental heat, fans, and on/off cycling. The higher the HSPF, the more efficient the heat pump will be. A heat pump with an HSPF rating of 6.9 will have an average COP of 2 for the heating season. To make this calculation, divide the HSPF rating by 3.4 to get the average COP.

Do not assume that the HSPF rating is an accurate predictor of the actual installed conditions. Again, depending on the part of the country in which you live, these ratings will be different. As an example, I live in Oregon. Using a heat pump west of the Cascade mountain range, with a well-designed system, you should see your rating close to the HSPF rating for your unit. However, if you live on the east side of the Cascades, where the temperature is much lower in the winter months, you will have an HSPF rating lower than the calculations.

You should attempt to find a heat pump that will give you an HSPF rating of at least 6.8. Some units can be rated as high as 9, but these units are very expensive. You will need to factor in the cost of the unit verses the total energy savings that the unit will provide.

SEER

The cooling performance of your heat pump is rated using the *seasonal energy efficiency rating* (SEER). The higher the SEER rating, the more efficiently the heat pump will cool your home. The SEER is the ratio of the amount of energy needed to remove the heat from a home versus the amount of energy needed to operate the heat pump. The SEER rating is normally much higher than the HSPF rating due to the fact that defrosting is not needed in the summer and you will not be using supplemental heat.

Unless you live in a part of the country where cooling is more important than heating, the HSPF rating will be more important that the SEER rating of your heat pump.

Defrost

Because there is a very cold liquid (refrigerant) flowing through the outdoor heat exchanger, ice can form on the coil just as it does in freezers. When the outdoor temperature gets below 40°F, the unit may have to defrost periodically. To do this, the heat pump must reverse the cycle to remove heat from the home to defrost the coil. This will cause a decrease in the efficiency of the unit.

Supplemental Heat

As the outside temperature gets colder, there is less heat to extract from the air, yet the home needs more heat to stay comfortable. At some outdoor temperature (known as the *balance point*), there will not be enough heat in the air for the heat pump to heat the home by itself. At this point, the unit will need supplemental heat to make up the difference between what the heat pump can or cannot supply and the needs of the home. In some units, there is an onboard electric resistance heating system with the heat pump. When the temperature reaches this balance point, the electric resistance heat will kick in. In some cases, where there is not a backup heating system on the heat pump itself, the unit will call for heat from your indoor heating system to make up the difference. This also will be the case if there is a failure in the heat pump for any reason. The backup or supplemental heating system will be used if the heat pump fails to operate.

Balance Point

When you plot the heating requirements for a home, it will look something like the graph in Fig. 26.2. As the outside temperature drops, the amount of heat that is available for the heat pump to extract from the air also drops. At the same time, the amount of heat that is required to keep the home at the desired temperature increases. At some point (in this example 31°F) the heat pump output and the home heating requirements match. This temperature is called the *balance point*. Anything below this temperature and supplemental heating will be required.

The capacity of the heat pump should be "sized" to match the desired balance point. In an area where the main consideration is heating, sizing may be a compromise between heating and cooling requirements. The chosen balance point will vary depending on the climate in which you live. Typical balance points should be between 27 and 35°F.

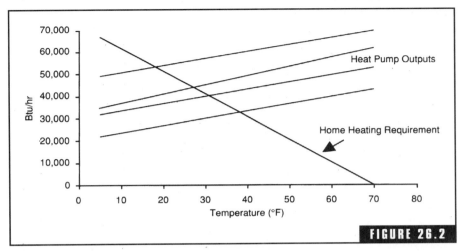

FIGURE 26.2

Balance point chart.

Btus

Btu is the abbreviation for British thermal unit. This is a small amount of energy, roughly the amount of energy given off by burning a wooden match. When we are talking about heat pumps, we talk in terms of tons. So what is a ton? Well, a ton is equal to 12,000 Btus of heat output when the air is 47°F or 12,000 Btus of cooling with the air is 95°F.

Now that I have introduced you to the different terms used in determining the efficiency of a heat pump, let's talk about how a unit is sized and the different ways to make a unit as energy efficient as possible.

Sizing the Unit

As a consumer, it will be hard for you to tell if the contractor you have selected has properly sized the unit to meet your needs. However, you can let the contractor know that certain criteria must be met. I will give you some of the most widely accepted criteria for determining if your new heat pump is sized properly for your home.

There are three reasons that a heat pump should be properly sized for your home:

1. *Cost.* Large equipment is more expensive than small equipment. If your heat pump is too large, you will spend more money for the equipment. If it is too small, your backup heating system will run more often, causing your electric bill to increase.

2. *Durability.* Most wear and tear on the compressor occurs at startup. Oversizing your heat pump will cause the equipment to cycle on and off more frequently. This will cause the compressor to wear more quickly and can lead to premature wear on this unit. A compressor is very expensive to replace, so you need to make sure that you do not install a unit that is too large for your home.

3. *Efficiency.* Oversized equipment has a shorter on cycle. This means that the unit will spend more time starting, which is the greatest cause of an inefficiently operating unit. It takes more energy to start a unit than when it is operating. The compressor has its largest amperage draw during the start cycle than at any other part of the cycle.

Heating and Cooling Load Calculations

Load Calculations

The only way to properly size a unit for your home is to have the contractor perform a heating and cooling load calculation on your home. This calculation should consider the following criteria:

1. The dimensions of your floors, basement walls, above-ground walls, windows, doors, and ceilings.

2. The amount of insulation value in these components. (Do you have single- or double-pane windows? What is the *R* value of the insulation in the walls and ceilings?, etc.)

3. Local weather conditions. Loads should be calculated for a cold winter day (but not the coldest) and a warm summer day (but not the hottest).

If you are going to be using the heat pump for heating purposes only, you will only have the contractor do the load calculations for the heat load. Conversely, if you are only going to use the unit for cooling, you only need to have the calculations done for cooling. However, using a unit such as a heat pump to perform only half the duties it is designed to do could be a waste of money for you as a homeowner. You should consult with your heating contractor for a more efficient way to meet your needs.

If you are going to have a heat pump installed, you will want to talk to the contractor to make sure that he or she has done these load calculations and that he or she is not simply guessing at the needs you have. Ask the questions about the calculated load to make sure that the contractor is sizing the unit properly for your home. This unit will be with you for a long while, and you want to make sure that you are being as energy efficient as possible to save you the most amount of money possible. Since there is an average return on investment (ROI) of 3 to 5 years on a heat pump, you can see that by not properly sizing the unit to your needs, the ROI can take much longer to accrue.

There are many ways to calculate load. The most common is the one developed by the American Society of Heating, Cooling, and Air Conditioning Engineers (ASHRAE, pronounced "ash ray"). This calculation is quite complicated and has been simplified by some contractors to speed up the process of determining this load.

There is a simple way for the average homeowner to do some quick calculations to check the contractor. This will give you a "rough" calculation that can help to see if your numbers are close to the contractors.

The first thing that you need to know is the total square footage of your home. This is done by measuring the outside of the home. In my case, I live in a manufactured home in Oregon. The size of my home is 28×56 ft. This gives me a total square footage of 1568 ft^2 (28 ft wide times the length of 56 ft). This will give you a rough number to start with. These calculations are done to estimate the heat gain in your home. Your contractor may use a different method to come up with these numbers, but this will give you some idea if the contractor's numbers come close to yours.

It is generally considered acceptable to assume that 1 ton of refrigerant is needed for every 600 ft^2 of living space. So, for our example, with a home that has 1568 ft^2 of living space, how many tons will be needed? If you said 2$^1/_2$ tons, you would be correct. As this example shows, the calculations do not come out even. I could consider going to the next larger size heat pump that would work for an 1800 ft² home, but this would be much larger that what is needed for my home. Thus 2¹/₂ tons will work for my home because this is enough to work with a 1500 ft² home. As I stated, the contractor will use more information to come up with the calculations, but this will, at least, give you an idea of what the numbers should look like.

Figures 26.3 and 26.4 show examples of the heat load and cooling load forms used to make these calculations.

Ways to Improve Your Heat Gain Ratio

There are several ways that a homeowner can improve the heat gain in his or her home. This is most important in the summer months when you are attempting to use your heat pump to cool your home. Some of the ways that you can use to improve the heat gain ratio are as follows:

1. *Windows.* Make sure that you have good windows in your home. They should be at least double-pane windows. If these windows face the sun, you can reduce your heat gain by keeping them closed, and close the blinds or curtains during the hot part of the day. Also make sure that there are no gaps between the window and the home. Caulk all openings to reduce the amount of draft that is coming in around the windows.

2. *Doors.* The best type of doors are solid core or steel. These doors have the best insulating values and will help to keep the heat out of the home. You also can improve heat gain by the addition of storm doors on the outside of the home. You also want to make sure that there are no drafts coming from the doors. You should check the weather stripping on the doors to reduce the amount of draft that comes in. The more draft there is around doors and windows, the more heat that is entering the home in the summer and cold air in the winter.

3. *Insulation.* You should have your home insulated if this has not been done already. This includes the walls, floors, and ceilings. The better insulated the home is, the easier it will be to cool the home in the summer.

WALLS: (Linear Feet)   °F, TEMP. DIFFERENCE

WALLS: (Linear Feet)	
2" Insulation	Lin. Ft. × 1.6
Average	Lin. Ft. × 2.6
WINDOWS & DOORS: (Area, Sq. Ft.)	
Single Glass:	Sq. Ft. × 1.13
Double Glass:	Sq. Ft. × 0.61
INFILTRATION - WINDOWS & DOORS: AVG.	Lin. Ft. × 1.0
Loose	Lin. Ft. × 2.0
CEILING: (Area, Sq. Ft.)	
Insulated (6")	Sq. Ft. × 0.07
Insulated (2")	Sq. Ft. × 0.10
Bulit-up Roof (2" insulated)	Sq. Ft. × 0.10
Built-up Roof (1/2" insulated)	Sq. Ft. × 0.20
No Insulation	Sq. Ft. × 0.33
FLOOR: (area, Sq. Ft.)	
Above Vented CrawlSpace	
Insulated (1")	Sq. Ft. × 0.20
Uninsulated	Sq. Ft. × 0.50
*Slab on ground	Lin. Ft. × 1.70
1" Perimeter Insulation	Lin. Ft. × 1.00

*Based on Liner Feet of outside wall   TOTAL HEAT   LOSS PER °F   BTU/HR/°F

Multiply total BTU/HR/°F x 30 and plot on graph below at 40°F. Draw straight line from 70 base point thru point plotted at 40°F. Intersection of this heat loss line with unit capacity line represents the winter design temperature in which the unit will heat the calculated space.

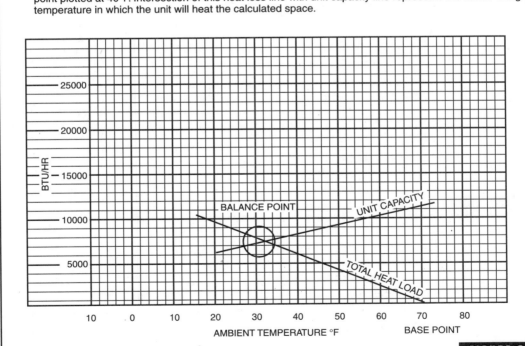

Heat load form.

The heat load form may be used by servicing personnel to determine the heat loss of a conditioned space and the ambient winter design temperatures in which the unit will heat the calculated space.

The upper half of the form is for computing the heat loss of the space to be conditioned. It is necessary only to insert the proper measurements on the lines provided and multiply by the given factors, then add this result for the total heat loss in BTU/Hr./°F.

The BTU/Hr. per°F temperature difference is the 70°F inside winter designed temperature minus the lowest outdoor ambient winter temperature of the area where the unit is installed. This temperature difference is used as the multiplier when calculating the heat loss.

The graph shows the following:

Left Hand Scale	Unit capacity BTU/Hr. or heat loss BTU/Hr.
Bottom Scale	Outdoor ambient temperature, bass point.
Heat Pump Model	BTU/Hr. capacity heat pump deliver at outdoor temperatures.
Balance Point	Maximum BTU/Hr. heat pump will deliver at indicated ambient temperature.

Below is an example using the heat load form:

A space to be conditioned is part of a house geographically located in an area where the lowest outdoor ambient winter temperature is 40°F. The calculated heat loss is 184 BTU/Hr./°F.

Subtract 40°F (lowest outdoor ambient temperature for the geographical location) from 70°F (Inside design temperature of the unit) for a difference of 30°F. Multiply 184 by 30 for a 5500 BTU/Hr. total heat loss for the calculated space.

On the graph, plot the base point (70°) and a point on the 40°F line where it intersects with the 5500 BTU/Hr. line on the left scale. Draw a straight line from the base point 70 through the point plotted at 40°F. This is the total heat load line.

Knowing that we have a 5500 BTU/Hr. heat loss, and we expect that our heat pump will maintain a 70°F inside temperature at 40°F outdoor ambient, we plot the selected unit capacity BTU/Hr. of the unit between 35° and 60° on the graph and draw a straight line between these points. Where the total heat loss line and the unit capacity line intersect, read down to the outdoor ambient temperature scale and find that this unit will deliver the required BTU/Hr. capacity to approximately 30°F.

FIGURE 26.3

Heat load form (*continued*).

Other factors to consider with heat gain are the number of people that you have in the home, the types of appliances you are using that produce heat, and the type of lighting that is used. All these items produce heat that must be removed from the home in the summer to properly cool the home.

These same items can be the cause of heat loss in the winter. If the home is not properly insulated or there are drafts around the doors and

HEAT GAIN FROM	QUANTITY	FACTORS DAY				BTU/Hr (Quality x Factor)

1. WINDOWS: Heat gain from sun		No Shades*	Inside Shades*	Outside Shades*	(Area x Factor)	
Northeast	___ sq ft	60	25	20 ___	Use	___
East	___ sq ft	60	40	25 ___	only	___
Southeast	___ sq ft	75	30	20 ___	the	___
South	___ sq ft	75	35	20 ___	largest	___
Southwest	___ sq ft	110	45	30 ___	load.	___
West	___ sq ft	150	55	45 ___	Use	___
Northwest	___ sq ft	120	50	35 ___	only	___
North	___ sq ft	0	0	0 ___	one.	___

* These factors are for single glass only. For glass block, multiply the above factors by 0.5; for double glass or storm windows, multiply the above factors by 0.8.

2. Windows: Heat gain by conduction (Total of all windows)			
Single glass	___ sq ft	14	___
Double glass or glass block	___ sq ft	7	___

3. WALLS: (Based on linear feet of wall)		Light Construction	Heavy Construction	
a. Outside walls				
North exposure	___ ft	30	20	___
Other than North exposure	___ ft	60	30	___
b. Inside Walls (between conditioned and uncond-itioned spaces only)	___ ft	30		___

4. ROOF OR CEILING: (Use one only)			
a. Roof, uninsulated	___ sq ft	1?	___
b. Roof, 1 inch or more insulation	___ sq ft	8	___
c. Ceiling, occupied spaces above	___ sq ft	3	___
d. Ceiling, insulated with attic space above	___ sq ft	5	___
e. Ceiling, uninsulated, with attic space above	___ sq ft	12	___

5. FLOOR: (Disregard if floor is directly on ground or over basement	___ sq ft	3	___

6. NUMBER OF PEOPLE:	___	800	___

7. LIGHTS AND ELECTRICAL EQUIPMENT IN USE	___ watts	3	___

8. DOORS AND ARCHES CONTINUOUSLY OPENED TO UNCONDITIONED SPACE: (Linear feet of width)	___ ft	300	___

9. SUB-TOTAL	xxxxxx	xxxxxxx	

10. TOTAL COOLING LOAD: (BTU per hour to be used for selection of room air conditioner).) _____ (Item 0) x _____ (Factor from Map) = ___

FIGURE 26.4

Cooling load form.

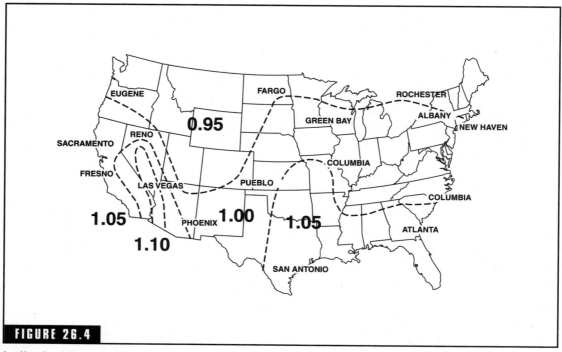

FIGURE 26.4

Cooling load form (*continued*).

windows, heat will escape, and the cost to heat the home will go up. Figure 26.5 shows some common heat loss points.

Other items to consider are insulation and sealing of the ductwork and changing the filters in the heating system every 3 months. One more item to consider when we are talking about reducing heat gain in the summer is skylights. I have these in my home, and I can tell you that they allow a lot of heat to enter the home. There are a couple of ways to reduce the heat gain produced by these skylights:

1. *Cover the skylights.* This can be done by placing a heavy blanket over them or constructing a box that goes over them on the roof. These items can be placed over the skylight when the days are going to be very hot. You also can remove them when the daytime temperature will be moderate.

2. *Blinds.* Several companies make blinds that are installed on the inside of the home so that the skylights can be closed off when they are not needed or when the temperature will be high during the day.

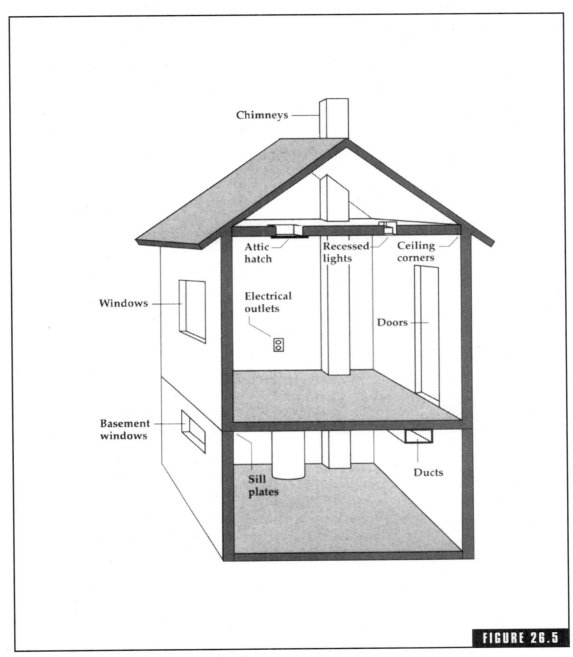

Common air leakage points. (*Courtesy of EREC.*)

Both these methods work to reduce the amount of heat that enters the home from these units.

There are several good sources of information that can show you the best way to conserve energy. One great source of information is

Energy Efficiency and Renewable Energy Clearinghouse
P.O. Box 3048
Merrfield, VA 22116
E-mail: doe.erec@inclinc.com
Phone: 1-800-273-2957

Operation and Maintenance of Heat Pumps

This chapter covers the operation and maintenance of heat pumps. I go into greater detail on how heat pumps work and the proper way that a homeowner can maintain them.

Heat pumps operate on the simple idea that they can remove the heat from a source, and this heat can then be used to heat or cool a home. When the temperature reaches a point where no more heat can be extracted, then a backup heat source is needed to maintain the temperature in the home. This is the basic principle of all heat pumps, regardless of their source of heat. In summer, this process is reversed so as to remove the heat from the home.

It is the movement of the refrigerant through these components that allows the heat pump to heat and cool a home. During the heating cycle, air blows across the outdoor evaporator coil that contains low-pressure refrigerant. This refrigerant has a very low boiling point. As

QUICK»TIP

Six basic components of a heat pump:

- Compressor
- Expansion valve
- Reversing valve
- Two heat exchangers
- Condenser
- Evaporator

the pump's fan blows air across the evaporator coil, the heat from the air is absorbed into the refrigerant, causing it to change from a liquid to a vapor. This vapor is then run through the compressor and is compressed. This causes the vapor to become a much hotter vapor as a result of the compression, and it is pumped to the indoor heat exchanger, also known as the *condenser.*

As the vapor moves through the condenser, the indoor blower blows air across the condenser to remove the heat from the vapor. As this vapor cools, it turns back into a liquid. The heat that is removed from the hot vapor is blown through the ductwork into the home as heat. The cooled liquid is then pumped through an expansion valve, where its temperature and pressure are reduced. Now it is a cooled liquid at low pressure, and it is again pumped through the outdoor heat exchanger to begin the cycle over again.

The cooler medium from which the heat is extracted is called the *heat source,* and the warm medium it is released to is called the *heat sink.* This is why heat pumps are usually called by their source and sink such as air-to-air, water-to-air, and ground-to-air.

The example above would be a typical cycle of an air-to-air heat pump, where the source of the heat is the air and the sink is also the air.

Maintaining air-to-air heat pumps is fairly simple. First, you want to make sure that there is clearance between the outside unit and any bushes or landscaping that is in place. The unit must be able to draw the air in from the outside to operate properly. Make sure that you keep all obstructions away from the unit. The evaporator coil must be kept clean. You can use a hose to clean the dust and dirt from this unit. Filters on the air movement unit must be changed every month so that the blower can move the maximum amount of air. It does not matter if this is a forced air heating system or an air mover, the need to change these filters is the same.

Water source heat pumps operate a little differently from air source heat pumps. Since water in ponds, wells, and springs ranges in temperature from 50 to 70° year round, they are a great source of heat.

A water source heat pump may use either a closed- or open-loop configuration. In the open-loop system, the water is pulled through an open pipe to the heat exchanger. Once the heat is extracted from the water, the water is then pumped to a disposal body of water. This system works when the water source is a well. If you are going to use a lake or pond as the source of water, there may be some environmental issues with regard to dumping the water back into the pond or lake. This is where a closed-loop system will work.

In a closed-loop system, the water in the pipes is circulated between the source and the heat exchanger and then back to the source. The closed-loop system eliminates the need for great amounts of water because the water in the pipes is used over and over again.

If you are considering this type of heat pump to heat and cool your home, you must first check your local and state laws in see if it is legal for you to use ground water as a heat source. In some areas, it is not legal to reintroduce water back into the ground after it has been used.

If there are no legal problems in your area, the next consideration is flow. A water source heat pump requires a flow of between 1.5 and 3 gallons per minute (gpm) per ton of heat pump capacity. This means that on a cold winter day, your heat pump could require as much as 14,000 gallons of water to maintain a comfortable temperature in the home. If you are considering using an open-loop water source, you need to consult with a qualified well-drilling operation to make sure that you have the capacity to operate such a heat pump.

Unless you have good water flow, you will need to drill two wells for the heat pump: one well for the water source and the other for dumping the return water. Cost must be considered in such a case, since the cost of drilling these wells will run between $5 and $15 per foot. Depending on the area where you live, the depth of the wells that must be drilled to get the proper flow will be variable. Additionally, the wells for the supply and return should be no closer than 100 ft so that the water being returned will not cool the water that is being used as the source.

If you cannot get the proper flow rate to operate a heat pump for the tonnage that is required for your home, you might consider operating two heat pumps in tandem. The temperature drop between these two units will only be about 10°F, so this will not adversely affect their operation. Cost must be considered in this case, because the cost to install the units will be more than that for a single unit, but this may be one alternative in the case of low water flow.

Maintenance of the open type of water source heat pumps is also a consideration versus a closed-loop system. Since ground water contains microorganisms, minerals, and sometimes sediment that can clog the water to refrigerant lines, you must have your water evaluated to determine whether it contains chemicals that may cause scaling, corrosion, or incrustation. Incrustation reduces heat transfer because it acts as an insulator. This also will reduce the efficiency of the pump. If your water source contains these types of minerals and deposits, you will need to consider the cost of treating the water. This treatment should only be done by trained professionals, since there can be dangerous fumes associated with treating the water. You also may have to consider installing a filter on the inlet side of the suction line. This can help to maintain the efficiency of the unit. If you find that you cannot get the proper flow rate required for an open-loop system, then you may want to consider changing to a vertical or horizontal closed-loop system.

Another consideration is the proper sizing of the pump. The pump must be large enough to supply the maximum amount of water to efficiently heat and cool the home while overcoming the friction losses of the pipe and heat pump. The pipes should be large enough to supply a flow rate of 10 gpm for every 36,000 Btus of heating capacity. These pipes must be buried below the frost line. This level will vary from a few inches to several feet depending on the part of the country in which you live.

The bottom line is that water source heat pumps are more efficient than air source heat pumps because of the fact that you are dealing with a heat source that is more constant than air. With proper installation and maintenance, you should realize a payback on this type of heat pump within 3 years.

Ground source heat pumps are another excellent way to heat and cool your home.

There are over 50,000 ground source heat pumps in operation in the United States today according to the Department of Energy. Since the ground, like water, maintains a constant temperature below the frost line, this is another good source of heat. This type of system yields a COP of 2.5 or more, on average. When the unit is operating in the cooling mode, the excess heat can be returned to the ground.

Ground source heat pumps use a series of pipes that are buried below the frost line. The configuration of the pipes can be either vertical or horizontal. Most of the heat pumps in use today are of the vertical type because they require much less land than the horizontal type. Both types use a closed-loop system to transfer the liquid to the heat exchanger. In the vertical configuration, moist soil will help the heat transfer process.

In a horizontal configuration, the depth of the pipes will depend on the climate and on the characteristics of the heat pump. The piping should be covered with sand and earth. Water, brine, or water treated with antifreeze circulates through the pipes, absorbing the heat from the ground. The amount of ground needed for this type of system to operate properly is roughly equal to the square footage of the house but never less than half that number. The length of piping required will depend on a number of factors, such as the location of the house, the quality of the construction, the moisture content of the soil, the heating and cooling loads, and the size of the heat pump. If possible, attempt to place the pipes under a garden. In the winter, the dark surface of the soil is exposed to the sun, aiding in solar absorption. In the summer, the garden will shade the ground and will help the ground to absorb the heat transfer from the home.

In a vertical configuration, the pipes will require about 1 to 2 ft^2 of surface area and a 150- to 400-ft vertical hole per ground coil set. Up to four pipes can be installed per set. Rock is more difficult to drill than soil, but rock is a better conductor of heat than soil. Soil has a tendency to collapse during the drilling process, but a good drilling company should have little trouble drilling to this depth.

The cost to install one of these units will depend on the soil conditions. The total cost to install one of these units, considering that you are using a non-forced air heating system, will be between $5000 and $8000. The lower cost is due to the aggressive marketing efforts of the local electric companies and new ground heat exchanger installation equipment that automatically places and backfills multiple pipes in a single trench. As with all estimates, the exact cost will vary depending on a number of factors.

Maintenance of these units is fairly simple. All that is required is that the filters on the air exchanger be changed every month for best airflow. The blower motor should be checked for proper lubrication, and the fins on the blower need to be cleaned when they become dirty.

As you can see, careful consideration must be given to determining which system will work best for you. Water and ground source heat pumps are more efficient than air-to-air heat pumps, but they have a much higher cost to purchase and install. Air-to-air systems do not require the amount of room to operate as the other types of heat pumps do and they are much less expensive to install. The choice of which unit is best depends on your needs and where you live. In any case, you will find that the amount of energy that you are currently using will decrease with the addition of a heat pump.

Energy is a precious commodity that we must all help to conserve. By adding a heat pump to your existing heating system, you will be doing your part to help conserve this resource.

Troubleshooting Heat Pumps

As with all the other forced air heating systems in this book, the heat pump can and will fail. Using a systematic approach to solving the problem is the best way to find and solve the problem quickly and efficiently.

Heat pumps have some rather unique characteristics that do not apply to other forced air heating systems. This unit is subjected to the weather, it contains pumps and coils that the normal forced air heating system does not have, and it can be used to heat or cool the home. However, there are some areas that the heat pump does have in common with other forced air heating systems. Some of these would be the thermostat, blower, fuses, and electricity used to power the unit. The skills needed to troubleshoot these units can be learned with practice. This chapter introduces you to these skills, and with practice, you will become as skilled a troubleshooter of heat pumps as you are with forced air heating systems. Let's begin with air-to-air heat pump troubleshooting, and then move to the water source heat pumps.

The Unit Will Not Operate

As with the other forced air heating systems, this call probably will be the most common call you receive. When a heat pump does not operate, the cause can be as simple as a blown fuse. One quick way to check for this is to set the blower on the air exchanger of the furnace to the "manual" setting. With a heat pump installed, the thermostat is the best place to go, since the thermostat will have a lever on it so you can check this setting. Move the lever to the "manual" setting and see if the blower comes on. You also should check to make sure that the lever for the heat pump is in the proper position. If you are making this service call in the summer, make sure that the lever is in the proper position so that the unit is calling for cooling. If this call is being made in the winter, make sure that the lever is set to call for heat.

If all the settings are correct and set and the blower does not come on, check for a blown fuse or tripped breaker. This will be located in the fuse box or breaker panel. If the panel is marked, check the breaker or fuse to see if it is tripped or blown. If so, replace the fuse or reset the breaker to make the blower come on. You also will need to check the disconnect on the outside unit to see if the lever got moved. If this solves the problem, you are finished. Check the operation of the unit while you are there to make sure that this was the cause of the problem. If this does not solve the whole problem, move to the outside unit.

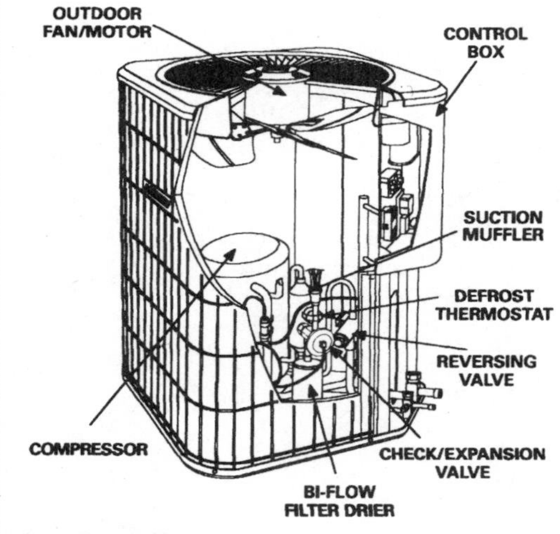

Refer to Fig. 28.1 for component location reference. You will want to check to make sure that you have power to the outside unit now that you have determined that you have power to the air mover. Close the disconnect so you can open the cover. Using your meter, check for power at the disconnect. If you have power to one side but not the other, one of the fuses in the disconnect is blown. Care must be taken when removing these fuses because you are dealing with high voltage. These

FIGURE 28.1

Heat pump components. (*Courtesy of Lennox Industries, Inc.*)

OUTDOOR FAN/MOTOR, CONTROL BOX, SUCTION MUFFLER, DEFROST THERMOSTAT, REVERSING VALVE, CHECK/EXPANSION VALVE, BI-FLOW FILTER DRIER, COMPRESSOR fuses are best removed with a fuse puller. Remove the fuses, and determine which one is blown. This can be done by taking a continuity reading on each one. If you do not get a reading, you have the blown fuse.

Replace all blown fuses, and restart the system to make sure that it is operating properly. If the system is operating properly and you do not see a reason for the fuse to have blown (e.g., bad motor, bad compressor, etc.), run the unit for a period of time to make sure that all is well. If it is, the problem is solved.

If you still do not have the unit operating, check the transformer. If you do not have any voltage reading on the output side of the transformer, this will need to be replaced. Turn the power off to the unit, and remove the screws that hold the transformer in place. Disconnect the wires from the back of the transformer, and remove the unit. Replace the transformer with one of the same output rating. Reconnect the wires, and turn the power back on to the unit. If this was the problem, check the operation of the unit to make sure that it is operating properly. If you have checked the power at all the locations and power is not the problem, you will need to go further into the system.

Liquid Pressure Too Low

If you suspect that the unit has a low liquid pressure, you will want to do some simple checks of the system before you hook up your gauges to check the actual pressure readings. One reason for a low liquid pressure could be that the coil is plugged. If the unit is located by large trees and is in an area where dust and dirt can get into the coil, you may have found the problem. Clean the coil with water or coil cleaning solution, and check the operation. It does not matter if the unit is attempting to heat or cool the home, the problems can be the same.

If the coil is not the problem with the low pressure and this is a new installation, make sure that the valves are installed properly in line. Check to make sure that the check valve and expansion valve were installed in the proper direction. Figure 28.2 shows the typical locations of these valves. If this is the problem, you will need to remove them and install them in the proper direction. Once this is done, start the unit and check the operation.

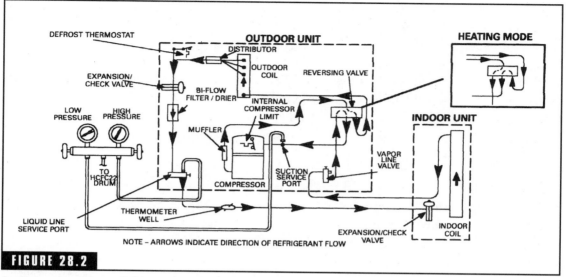

FIGURE 28.2

Cooling cycle with manifold connection. (*Courtesy of Lennox Industries, Inc.*)

The other problem that can cause low pressure is a blocked or restricted line. You need to check to make sure that there are no crushed lines. If you find a line that is crushed or bent, you will need to correct the problem before the unit will operate properly. Figures 28.3 and 28.4 show the proper installation of the lines.

Another cause for this problem is a leak in the line. Connect your gauge set to the unit, and check the pressure (Fig. 28.2). If you have determined by performing all the other checks of the system that the cause is lack of liquid in the lines, you will need to charge the system. More important than this is finding out why the charge is low. The most common cause is a leak in the system. You will need to do a leak check of the system to determine where the leak is and fix that problem before you charge the system or you will be wasting time and money. Use dye to find the cause of the leak, and fix the leak. Once this is done, charge the system and check the operation. If the system is operating properly, you have solved the problem.

If you have attempted all these suggestions and the failure points to the compressor, it will have to be replaced. I will not even attempt to instruct you on how to replace a compressor in this book. The best course of action is to refer to the manufacturer's service bulletin for

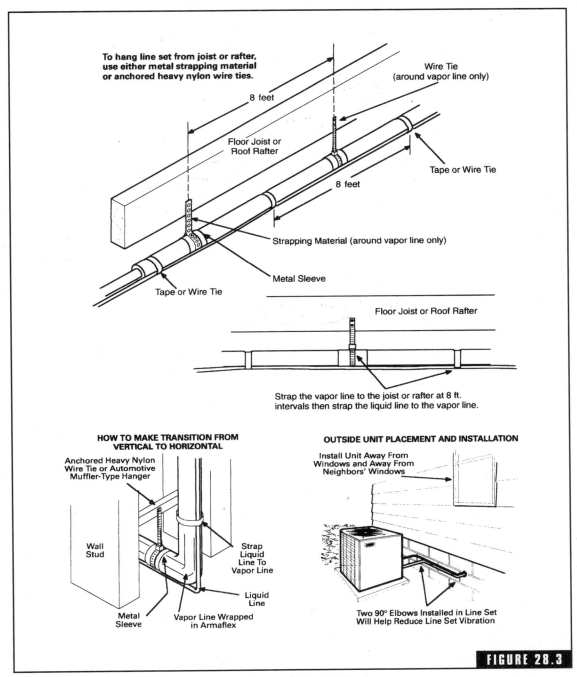

To hang line set from joist or rafter, use either metal strapping material or anchored heavy nylon wire ties.

8 feet

Floor Joist or Roof Rafter

Wire Tie (around vapor line only)

Tape or Wire Tie

8 feet

Strapping Material (around vapor line only)

Metal Sleeve

Tape or Wire Tie

Floor Joist or Roof Rafter

Strap the vapor line to the joist or rafter at 8 ft. intervals then strap the liquid line to the vapor line.

HOW TO MAKE TRANSITION FROM VERTICAL TO HORIZONTAL

Anchored Heavy Nylon Wire Tie or Automotive Muffler-Type Hanger

Wall Stud

Strap Liquid Line To Vapor Line

Liquid Line

Metal Sleeve

Vapor Line Wrapped in Armaflex

OUTSIDE UNIT PLACEMENT AND INSTALLATION

Install Unit Away From Windows and Away From Neighbors' Windows

Two 90° Elbows Installed in Line Set Will Help Reduce Line Set Vibration

FIGURE 28.3

Line installation—horizontal runs. (*Courtesy of Lennox Industries, Inc.*)

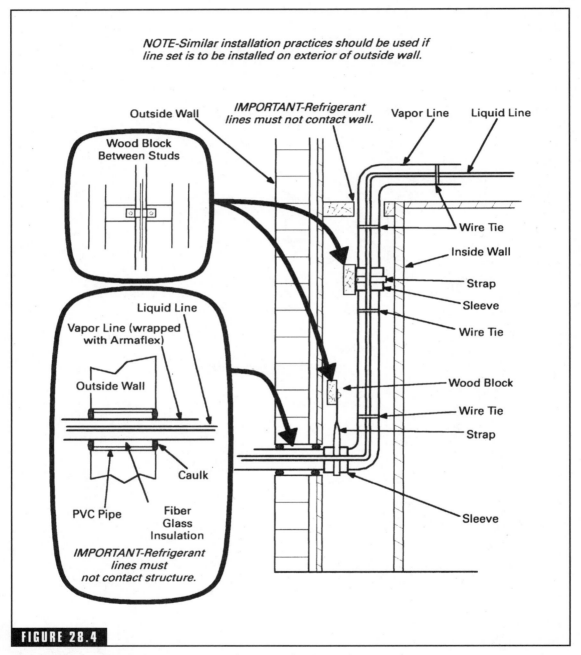

NOTE-Similar installation practices should be used if line set is to be installed on exterior of outside wall.

Wood Block Between Studs

Outside Wall

IMPORTANT-Refrigerant lines must not contact wall.

Vapor Line

Liquid Line

Wire Tie

Inside Wall

Strap

Sleeve

Wire Tie

Liquid Line

Vapor Line (wrapped with Armaflex)

Outside Wall

Wood Block

Wire Tie

Strap

Caulk

PVC Pipe

Fiber Glass Insulation

IMPORTANT-Refrigerant lines must not contact structure.

Sleeve

FIGURE 28.4

Line installation—vertical runs. (*Courtesy of Lennox Industries, Inc.*)

this information. If you are the homeowner, this procedure is not recommended for you. This requires special tools and equipment that you do not have. There are also environmental concerns to deal with in regard to the type of refrigerant that is used, and this is best left to the professional.

Liquid Pressure Too High

As with the low liquid pressure, there are some common causes for high liquid pressure as well. The most common of these is a plugged coil. If the coil becomes plugged, this will cause the unit to work harder, resulting in high liquid pressure. Inspect the coil for signs of dirt and debris. Clean the coil if this is found to be the problem. If the coil is found to be fine, turn your attention to the outdoor fan. Is the fan running? This fan should be running the whole time the unit is operating. If the fan is not operating, it will need to be checked and replaced if necessary. Figure 28.5 shows this operation. You also should check to make sure that the fan blades are secure to the motor shaft. If the blades are secure (this should only be checked with the power turned off) and the motor is not operating, check the capacitor. If the capacitor is not operating properly, it will not start the motor. Check this first before replacing the motor.

If the motor and coil are not the problem, check the lines for obstructions. If the lines are obstructed, this will cause the unit to have a problem.

The next item to check is the refrigerant. You want to check the amount of refrigerant in the lines. If the system is overcharged, you will need to remove some of the charge so that the unit will operate properly.

With the outside unit checked out, you can turn your attention to inside the home. Check the floor registers to make sure that they are open. Not getting enough heat or cooling can be caused by the registers

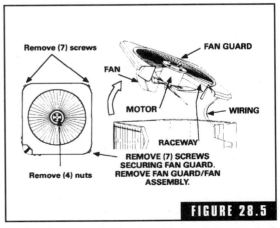

Outdoor fan removal. (*Courtesy of Lennox Industries, Inc.*)

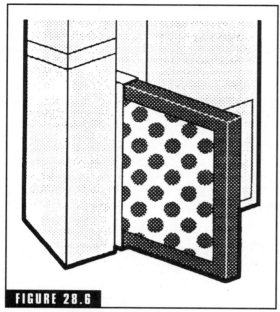

FIGURE 28.6

Filter installation. (*Courtesy of EREC.*)

being closed off. You also should check the condition of the filters in the heating unit. Remove and replace the filters if they are plugged. Figure 28.6 shows one such filter installation.

While you are at the blower, check the condition of this unit as well. Many causes can be traced to the blower being dirty and not being able to blow enough air.

The last item to check is the ductwork. If you have investigated all the other causes and you cannot find an answer, you may want to investigate the sizing of the ductwork. If the unit was added onto an existing heating system, the ductwork may be too small for the job. Have a heating and cooling professional do the calculations and make recommendations.

Pressures Are Normal, but There Is Still Not Enough Heat/Cooling

When you get a call to a home that is having a problem getting enough heat or cooling and you have checked all the pressures, you will need to check some uncommon causes. To start with, turn the power off to the unit, and inspect the electrical system. Check for worn or broken wires, for wires that are loose, and check the screws in the terminal blocks. If you do not find the problem there, check to make sure that all the ducts have air blowing from them. In some homes, the ductwork is not the standard sheet metal ductwork. In some cases, trunk lines are used to connect to the registers. Figure 28.7 shows different types of ductwork. These lines can be held in place with cable ties that come loose. You may have to inspect each of the lines to make sure they are connected properly. While you are inspecting the lines, check for leaks in the ductwork. Figure 28.8 shows potential air leak points. Any air that is leaking out of the ductwork into the basement or crawl space is not making its way into the home. Seal any leaks that you find.

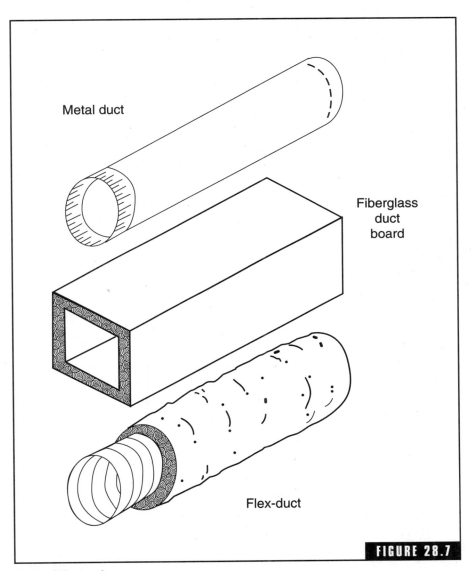

Metal duct

Fiberglass
duct
board

Flex-duct

FIGURE 28.7

Types of ductwork.

If the problem is not in the ductwork, check the filters and blower on the air mover or heating system. These units can only blow in as much air as they can draw into the blower. Dirty filters and cages on the blowers will cause the units to work much harder and produce less heat/cooling than is expected.

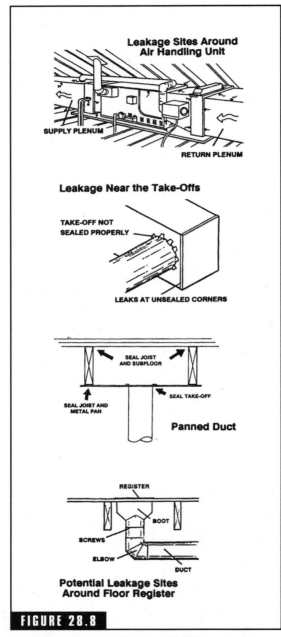

Leakage Sites Around
Air Handling Unit

SUPPLY PLENUM

RETURN PLENUM

Leakage Near the Take-Offs

TAKE-OFF NOT
SEALED PROPERLY

LEAKS AT UNSEALED CORNERS

SEAL JOIST
AND SUBFLOOR

SEAL TAKE-OFF

SEAL JOIST AND
METAL PAN

Panned Duct

REGISTER

BOOT

SCREWS

ELBOW

DUCT

**Potential Leakage Sites
Around Floor Register**

FIGURE 28.8

Sources of air leaks. (*Courtesy of EREC.*)

If the cause cannot be traced to these items, ask the homeowner if he or she has done any expansion to the home. If the unit was installed and sized for a certain amount of square feet and this has been expanded, this could be the cause. It also may be that the unit was not sized properly in the first place. You may remember the quick formula given in this section to determine the proper size. Use this formula to do a quick calculation of the current heat pump:

$$\text{Length} \times \text{width} = \text{total square feet of home}$$

Divide this number by 600 (the number of square feet 1 ton will cool) to get the size of the heat pump that should be installed. If the number is off, you may have discovered the problem. The only solution for this type of situation is to replace the unit with a larger one that will do the job.

The Unit Will Not Defrost

In cold climates, the unit will tend to develop frost. When this happens, the efficiency will drop. There is a cycle on the unit that will enable it to defrost to eliminate this problem. If the unit will not defrost, this can cause major problems.

Check the indoor coil and filters to make sure that they are in proper operating condition. Replace the filters if they are plugged, and clean the coil if needed.

If these two items are not the cause, check the reversing valve. During the reversing cycle, warm air from the home is used to defrost the coil. If the reversing valve is not operating properly, this will cause the problem. Check the relay on the reversing valve to make sure that it is operating properly. If not, replace the relay. There is also the possibility that the defrost control is defective. If it is defective, replace it. Figure 28.9 shows the location of the defrost control on one unit. Since every heat pump manufacturer is different, you may need to refer to the operations manual for your unit for more information.

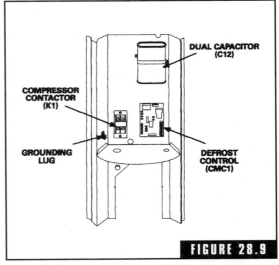

Control box with defrost control. (*Courtesy of Lennox Industries, Inc.*)

The Unit Will Not Stop Defrosting

If you have a unit that will not stop defrosting, you have a unit that is costing you a lot of money to operate. This is so because while the unit is using the indoor air to help defrost the coil, the backup heating system must make up for the lack of heat.

When you come across this type of situation, check the filters and coil inside the home first. These are simple to check and can be eliminated or solved quickly. If this is not the problem, check the reversing valve. There is a chance that this valve is stuck open or the relay has failed and is calling for the unit to continue to defrost.

Another cause of this problem is that the temperature bulb that controls the defrost cycle might be exposed to the elements and is calling for the unit to defrost. It also could be a defective defrost control.

If all these areas check out, check the charge in the system. If the charge is too low, this will cause the unit to freeze and the defrost cycle to run all the time, attempting to defrost the unit.

There can be other causes for failure, and only experience will teach you how to look for these other causes. This chapter was designed to give you some basic causes for failure and how to troubleshoot them. I have included two more illustrations at the end of this chapter. Figure 28.10 shows component location on the board, and Figure 28.11 shows the refrigeration components of a typical heat pump.

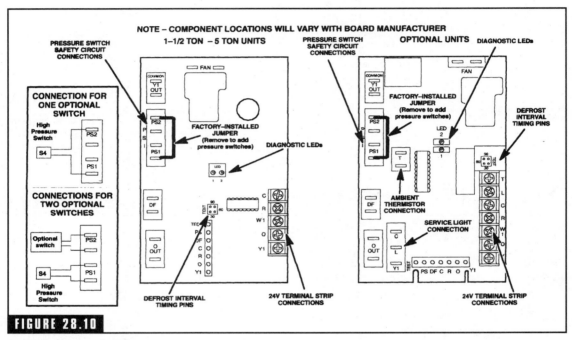

FIGURE 28.10

Board component location.

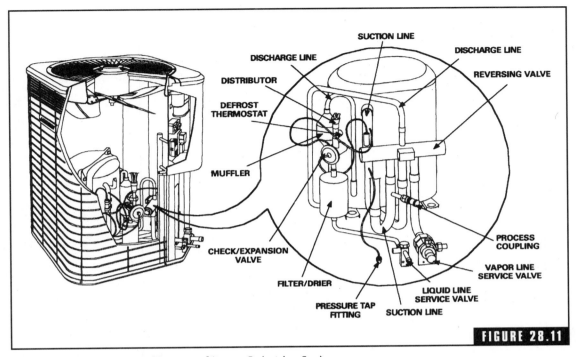

Refrigeration components. (*Courtesy of Lennox Industries, Inc.*)

FIGURE 28.11

Conclusion

As a heating service technician, there is one thing above all that you must be—and that is professional. You are being invited into someone's home to perform a service. You will be judged by the client on how efficient and professional you are. You must always remember that you are representing the company that you are working for. The impression that your leave with the client will either help to grow the business or could help to cause it to fail.

As a heating professional, you will be called on to service all types of forced air heating systems. Some of these installations will be new; some will be old. You need to understand how to service and maintain many different types of forced air heating systems from many manufacturers. You need to remember that when you are dealing with older systems, you must always perform the special checks of the system to make sure that the system is safe to operate. You are the front line of defense for the client to make sure that his or her forced air heating system will operate in a safe mode all winter long. You also must be able to efficiently troubleshoot and repair any problem that comes up during the heating season, and this is the reason this book was written for—to give you the knowledge to perform the checks and repairs that will keep your clients satisfied and allow them to judge you as the true professional that you are.

Some points to remember when you arrive at a client's home:

1. Present yourself in a professional manner. Wear clean uniform or clothes.

2. Address the client as Mr., Mrs., or Miss. Never use the client's first name.

3. Be organized. This means that you should have all the tools and equipment you need for the job. You should have a set of tools designed for the heating system that you will be working on.

4. Keep the area clean. Pick up all rags and debris you use at the job site.

5. Respect the client's home at all times.

6. Clean up any oil that you spill.

7. If the client is not home, leave the home in the same condition that you found it.

8. Always leave any parts that you replace so the client can see what was done.

By following this short list, you will leave the client with the impression that you and your company are true professionals.

Parts Inventory for a Well-Stocked Repair Truck

General tools

1. Tool box
2. Tool belt
3. Drop light
4. Flashlight
5. Torch
6. Rags
7. Oil pan
8. Oil can
9. Special tools as needed
10. Kitty litter for oil spills

11. Vacuum cleaner

12. Paint brush

13. Flaring tool

14. Tubing cutter

Gas heating supplies

1. Thermocouples

2. Fan controls

3. Gas valves

4. Aluminum tubing

5. Power pile generator

6. Pilot orifices

Oil supplies

1. Different sizes of nozzles

2. Oil filters

3. Burner motors

4. Different styles of oil pumps

5. Couplers

6. Different styles of electrodes

7. Electrode connectors

8. Ignition transformers

9. Cad cells

10. Cad-cell relays

11. Stack controls

Electrical supplies

1. Different styles of fan controls

2. Different styles of elements

3. Different styles of limit controls

Heat pump supplies

1. Condenser motor
2. Contractors
3. Defrost control
4. Compressor
5. Freon

Miscellaneous supplies

1. Transformers
2. Box filters
3. Filter media
4. Blower belts
5. Blower motors
6. Thermostats
7. Humidifier nozzles
8. Humidifier filters
9. Miscellaneous fittings
10. Copper tubing

Questions to Ask Before Going on a Service Call

1. What is the problem?

2. What type of heating system do you have?

3. Is the thermostat calling for heat?

4. Did you check all the fuses?

5. When did you last receive fuel (oil and propane units)?

6. Is the pilot lit (gas units)?

7. Did you press the reset button? How many times (oil units)?

8. Is your oil tank inside or outside?

9. Is the line insulated?

10. When was the last time the unit was serviced?

11. Can this service wait until morning (some shops charge more at night)?

12. What is your address and phone number?

Before leaving for a service call at night, check your book, if you have one, to see if the person calling is a COD customer. If so, call the person back, and explain that you will need to be paid in advance before you begin work. You will need to evaluate the repair, give an estimate, and collect the cash prior to beginning the work.

Oil Nozzle Selection

Btu of heating system	Nozzle needed
50,000	0.35 gph
55,000	0.40 gph
60,000	0.45 gph
65,000	0.50 gph
70,000	0.50 gph
75,000	0.55 gph
80,000	0.60 gph
85,000	0.60 gph
90,000	0.65 gph
95,000	0.70 gph
100,000	0.70 gph
110,000	0.80 gph
115,000	0.85 gph
120,000	0.85 gph

Btu of heating system	Nozzle needed
125,000	0.90 gph
130,000	0.90 gph
135,000	1.0 gph
140,000	1.0 gph

Note: For heating system Btu outputs not listed, divide the Btu output by 140,000 to get the gph of the nozzle needed. Check the nameplate to get the angle of spray needed for the nozzle size selected.

Sample of Summer Tune-Up Checklist

ABC HEATING COMPANY

DATE: _____   CUSTOMER NAME: _____

ADDRESS: _____

Gas heating system

___ Check thermostat

___ Check/clean pilot

___ Check/replace thermocouple

___ Clean/adjust burners

___ Check safety

___ Inspect heat exchanger

___ Oil blower motor

___ Inspect/replace belt

Oil heating system

___ Check thermostat

___ Replace oil filter

___ Bleed system

___ Check coupler

___ Check ignition transformer

___ Replace nozzle

___ Check/replace electrodes

___ Clean cad cell

Gas heating system	Oil heating system
__ Check/clean blower fins	__ Clean induction air cage
__ Check/replace filter	__ Inspect heat exchanger
__ Run check unit	__ Inspect oil blower motor
Electric heating unit	__ Inspect/replace belt
__ Check thermostat	__ Check/replace filters
__ Check/replace filter	__ Check safety
__ Oil blower motor	__ Run check unit
__ Check safety	Parts replaced/comments:
__ Run check unit	_____

Trouble Call Sheet Sample

ABC Heating Company

DATE: _____
CLIENT NAME: _____
ADDRESS: _____
TIME AT CALL: _____ TIME COMPLETED CALL: _____
_____

DESCRIPTION OF THE PROBLEM:

_____
_____
_____
_____
_____

ACTION TAKEN TO RESOLVE THE PROBLEM:

_____
_____
_____
_____
_____

SUPPLIES USED:

_____

_____

_____

_____

_____

_____

INDEX

ABOUT THE AUTHOR

Roger Vizi has over 15 years' experience in maintaining and repairing home heating systems and high-pressure steam boilers. He is currently senior buyer for SPM, an Oregon-based company that specializes in the manufacturer of plastic-injection molded parts. He is also the author of *The Homeowner's Guide to Basic Gas Furnace Repair*.